CNC SECRETS UNLOCKED

Advanced programming techniques and career boosting skills for machinists and engineers

RAJEEV M SREEDHARAN

Copyright

This book is dedicated to

my beloved wife,

Saritha Rajeev

Your quiet strength, unwavering support, and constant presence have been my greatest blessing. Like a shadow that never leaves my side, you have stood with me through every step of this journey.
Without you, this book would never have become a reality.

Acknowledgements

This book is the result of years of learning, experimentation, challenges, and growth. It would not have been possible without the support and encouragement of many wonderful individuals.

First and foremost, I express my deepest gratitude to my wife, **Saritha Rajeev**, whose unwavering support, patience, and belief in me have been the foundation of this journey. Her quiet strength, encouragement, and sacrifices made this dream possible.

To my family—thank you for your love and for always inspiring me to move forward with confidence and purpose.

I would like to make a special mention of my early colleague, **Mr. Mukund Sargar**, who found a teacher in me. Your trust and enthusiasm reinforced my belief in sharing knowledge and sparked my commitment to mentorship in the machining domain.

A heartfelt thanks also to my son, **Mr. Siddharth Rajeev**, whose enthusiasm, and support helped me set up my YouTube channel (CNC Learning - https://www.youtube.com/@CNCLearning) —the platform where my teaching journey truly began. Your belief in me and your help in embracing the digital world opened a new chapter in my life and learning.

I extend my sincere gratitude to **Mr. Gyan Narayan**, who hand-held me through this journey, guiding me step by step as I wrote and shaped this book. Your mentorship, encouragement, and support in the writing and publishing process gave me the confidence to bring this work to life.

To my mentors, colleagues, teachers, and industry experts—thank you for shaping my technical journey. Each interaction, challenge, and shared insight has contributed to the depth of this work.

To the global CNC and manufacturing community—machinists, engineers, and learners—your passion and dedication are the inspiration behind this book. This work is dedicated to everyone who strives for excellence through precision and continuous improvement.

Finally, to every reader—thank you for choosing to learn and grow with this book. May it support your journey, strengthen your skills, and inspire you to elevate the future of manufacturing.

— Rajeev M. Sreedharan

CNC Educator · 30+ Years in Manufacturing

About the Author

Rajeev M. Sreedharan is a manufacturing leader, CNC expert, and educator with more than 30 years of hands-on experience in precision machining and aerospace manufacturing. He has guided engineers, machinists, and organizations toward world-class machining practices, helping bridge the gap between shop-floor reality and advanced engineering principles.

Rajeev is passionate about demystifying CNC technology and empowering the next generation of machinists and manufacturing engineers. His journey as a teacher began early in his career when colleagues recognized his natural ability to simplify complex technical concepts. Today, through his YouTube channel CNC Learning, industry workshops, and technical writing, he continues to inspire professionals across India and beyond.

When he's not teaching or refining machining strategies, Rajeev enjoys learning new technologies, mentoring young talent, and contributing to India's mission of building world-class manufacturing capabilities.

FREE CNC Programming Mastery Kit

Test, Practice & Improve Your CNC Coding Skills

A professional practice & testing pack used by CNC programmers and instructors

What You'll Get — FREE

- Fanuc-style test templates
- G-code practice assignments
- Self-evaluation answer keys
- Ideal for CNC programmers, students & trainers

Bonus: Early access to CNC workshops + expert machining insights

Access Your CNC Mastery Kit

Scan the QR code

or visit:

https://rajeev-m-sreedharan.kit.com/2acdd9d795

Instant access. No spam. Unsubscribe anytime.

"Knowledge shared is knowledge multiplied. Let's build a smarter machining world together."

— Rajeev M. Sreedharan

CNC Educator · 30+ Years in Manufacturing

Table of Contents

Chapter 1: Foundations of CNC

CNC FOCUS POINTS

From Concept to Advanced Motion!

By completing this chapter, engineers and manufacturing professionals will gain mastery over the core principles, historical context, and technical elements about CNC.

1.1 Introduction — Revolutionizing Manufacturing

1.1.1 Definition and Significance of CNC

Computer Numerical Control (CNC) is a manufacturing process in which computers precisely direct machine-tool movements through digital instructions known as **G-codes**. This technology forms the operational bedrock of modern manufacturing—transforming CAD/CAM designs into tangible parts with exceptional accuracy, repeatability, and efficiency.

At its foundation, a CNC machine integrates:

- A **traditional machine tool** (such as a mill or lathe)
- A **CNC controller** (dedicated computer)
- A **spindle** (to hold the tool or workpiece)

- **Servo drives** (to execute program-controlled movements along multiple axes)

Understanding this concept of programmed motion is the first step toward mastering CNC technology.

1.1.2 Why CNC Matters

The adoption of CNC has transformed manufacturing by offering:

- **Repeatability:** Once validated, CNC programs guarantee identical results across countless cycles.
- **Speed & Efficiency:** Continuous, optimized operation with automated toolpaths and tool changes.
- **Complexity:** Multi-axis motion enables intricate features impossible with manual methods.
- **Scalability:** The same toolpath can serve prototypes, small batches, or mass production.
- **Applications:** Aerospace (turbine blades), Medical (implants), Automotive (engine blocks), Tooling, and Electronics.

1.2 The Core Technology

1.2.1 Closed-Loop Control System

The control system acts as the **nervous system** of a CNC machine, interpreting digital commands and translating them into precise, coordinated motion through continuous feedback.

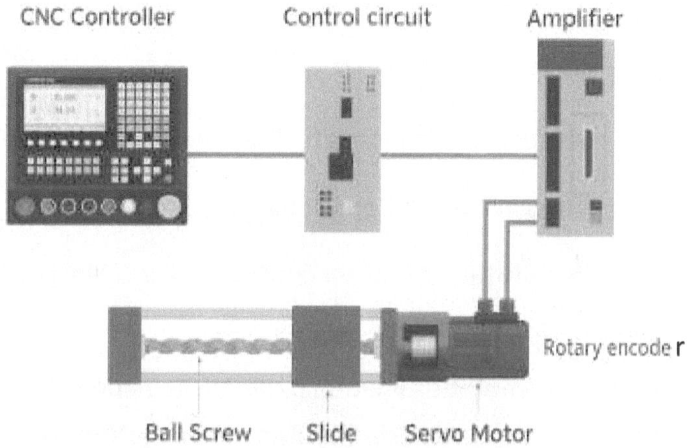

Signal flow within a closed-loop system:

1. **Command:** The controller reads the program and issues a movement command.
2. **Drive:** The signal is amplified to power the respective servo motor.
3. **Motion Translation:** Rotary motor motion is converted to precise linear movement via a **ball screw mechanism**.
4. **Feedback & Correction:** A rotary encoder monitors actual rotation and returns real-time data to the controller, which instantly compares and corrects any positional errors.

This self-correcting system enables tolerances measured in microns.

1.2.2 Precision Fundamentals

Three measures define CNC quality:

- **Accuracy:** Closeness to the intended dimension.
- **Repeatability:** Consistency across runs.
- **Resolution:** Smallest controllable movement increment.

Verification methods include cyclic testing and high-precision inspection using **CMMs**, **micrometers**, and **calipers**.

1.3 The Evolution of CNC Technology

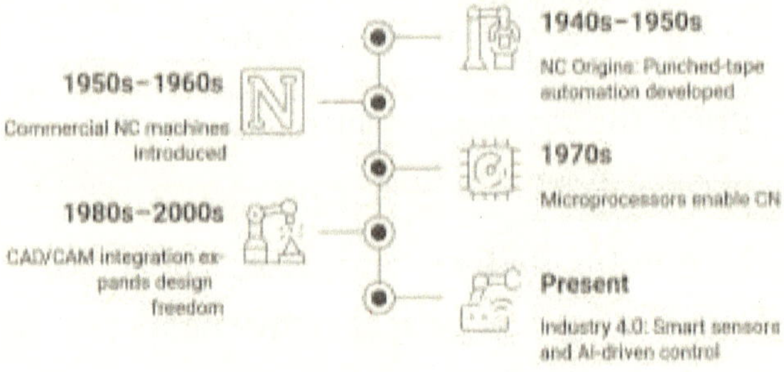

1.4 CNC Machine Anatomy and Classification

1.4.1 CNC Machine Anatomy

All CNC machines share these key elements:

- **Structure:** Base, columns, and bed/table for rigidity.
- **Spindle:** Controls tool speed and torque.
- **Drives:** Servo or linear motors with ball screws for positional control.
- **Coolant & Chip Management:** Ensure thermal stability and clean operation.
- **Controller:** Executes G/M-codes, plans motion, and manages tool life.

1.4.2 Specialized Machine Components

Machine Type	**Key Component**	**Function**
CNC Turning Center	**Chuck**	Rotating clamp that secures the workpiece.
CNC Turning Center	**Turret**	Indexes multiple tools rapidly into cutting position.
CNC Turning Center	**Tailstock**	Supports the free end of long workpieces along Z-axis.
CNC Milling Center	**Work Table**	Mounting surface moving in X-Y directions.

Machine Type	Key Component	Function
CNC Milling Center	**Tool Magazine**	Holds multiple cutting tools for automated changes.
CNC Milling Center	**Automatic Tool Changer (ATC)**	Swaps tools between magazine and spindle for automation.

1.4.3 Machine Classification

CNC machines are classified by process and axis count:

- **By Process:** Turning, Milling, Turn-Mill, and Mill-Turn.
- **By Axes:** 2 to 9 axes; five axes cover most industrial requirements.

1.5 Coordinate Systems and Axis Conventions

1.5.1 Cartesian Coordinates and Right-Hand Rule

The foundation of CNC motion is the **Cartesian coordinate system**.

For Vertical Machining Centers (VMCs):

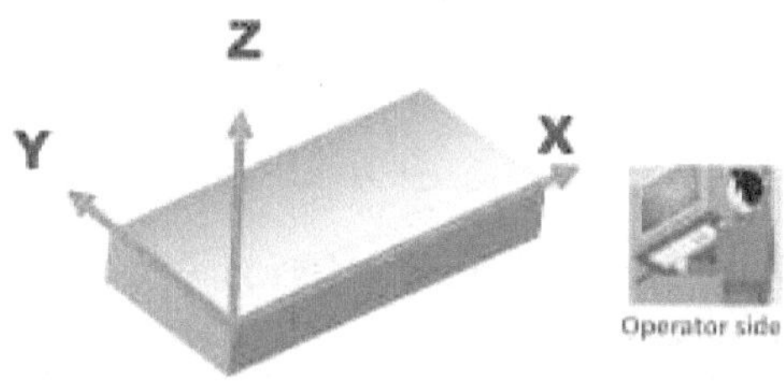

- **Z-Axis:** Parallel to spindle; positive upward.
- **X-Axis:** Horizontal (left-to-right); positive to the right.
- **Y-Axis:** Perpendicular to X and Z; positive away from the operator.
- **Rotary Axes (A/B/C):** Rotations around X/Y/Z respectively.

◆ **Key Principle:** Always visualize the tool moving around a fixed workpiece, even if the machine table physically moves.

1.5.2 Turning Center Axes

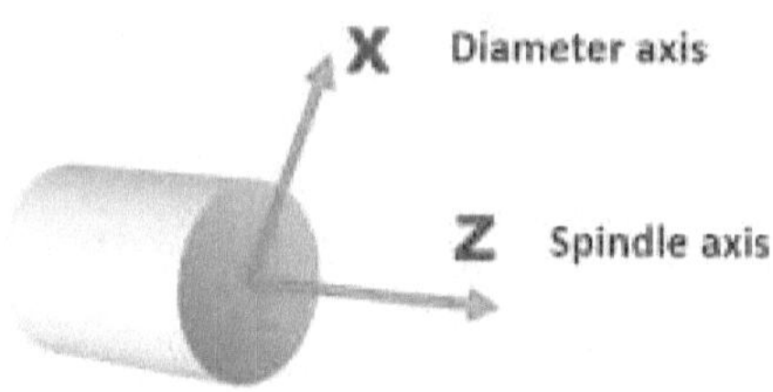

- **Z-Axis:** Parallel to spindle centerline; Z+ moves away from spindle.

- **X-Axis:** Perpendicular to spindle; X+ moves away from centerline (diameter direction).

1.5.3 Reference Frames and Offsets

- **Machine Coordinates (G53)** → Controller's absolute reference.
- **Work Offsets (G54–G59)** → User-defined zero locations.
- **Tool Offset (G43/H)** → Compensates tool length to ensure accurate depth.

1.6 Motion Strategies and Interpolation

1.6.1 Positioning Systems

- **Point-to-Point (PTP):** Rapid moves between fixed points (e.g., drilling). Controlled motion between points is not possible in PTP system.
  Example → G00 X10 Y10
- **Continuous Path:** Smooth, controlled cutting motion.
  Example → G01 X20 Y0 F500 (Linear) or G02 X20 Y0 I10 J0 (Arc).

1.6.2 Interpolation and Feed Control

- **G01:** Linear Interpolation
- **G02/G03:** Circular (Clockwise/Counterclockwise)
- **Helical:** Used for ramping or thread milling

Feed/Speed Modes:

- G96 → Constant Surface Speed
- G95 → Feed per Revolution
- G94 → Feed per Minute

1.7 Advanced Multi-Axis Machining

1.7.1 4-Axis Machining

Adds one rotary axis (A/B/C) in addition to three linear axes, allowing the workpiece to rotate for multi-side operations without re-clamping.

1.7.2 Hybrid Turn-Mill Centers

Combine turning and milling in one setup:

- **Turn-Mill Centers:** Primarily turning machines with light milling capability.

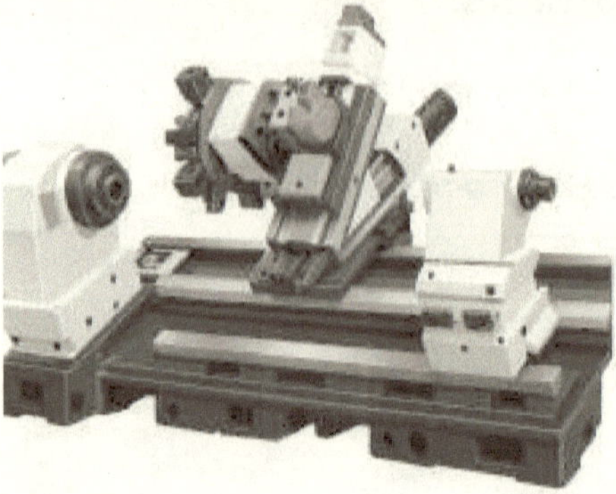

- **Mill-Turn Centers:** Perform both operations equally well; often include dual spindles. Unlike turn-mill centers, Mill-turn centers will have milling spindle, ATC, and tool magazine.

1.7.3 5-Axis Configurations

Enable simultaneous movement of five axes (3 linear + 2 rotary) to create complex free-form surfaces.

Common Architectures:

1. **Table–Table:**

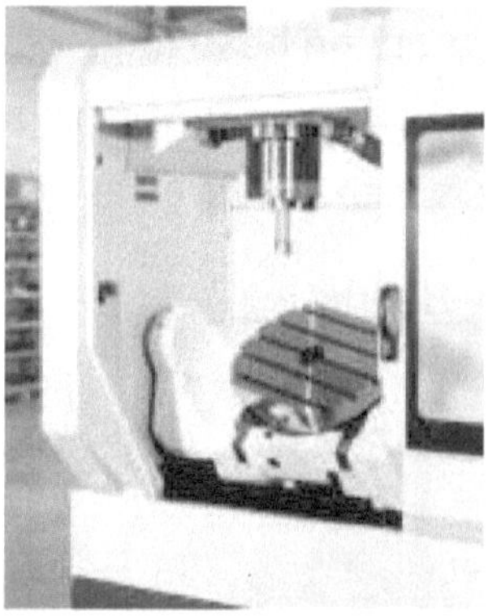

Both rotaries (A and C) on the table. Beneficial for parts with lesser weight or parts with sizes fit into the rotary table. Good for better chip evacuation.

2. **Head–Table:**

One rotary on head (e.g., B-axis tilt) and one on table (C-axis rotation). This configuration will allow for small as well as big parts.

3. **Head–Head:**

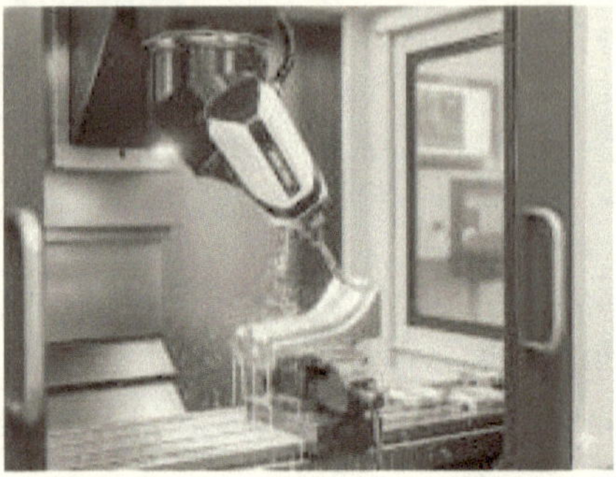

Both rotaries on the spindle head (B and C). This configuration is best suitable for heavy parts or long parts which is not suitable to rotate.

PRECISION RECAP

CNC (Computer Numerical Control) represents the perfect fusion of mechanical engineering and digital automation—transforming CAD/CAM designs into precise, repeatable, and efficient physical components. This chapter established the conceptual, structural, and operational foundation of CNC technology, illustrating how control systems, mechanics, and computation converge to achieve micron-level accuracy and versatility.

It also outlined the classification of CNC machines based on both process type and axis configuration. By process, machines are divided into Turning, Milling, Turn-Mill, and Mill-Turn systems, each optimized for specific manufacturing operations. By axis count, machines range from 2-axis basic configurations to advanced multi-axis systems—with 5-axis machines dominating modern manufacturing for their ability to create complex geometries in a single setup.

By connecting core principles such as coordinate geometry, servo feedback, machine architecture, and machine classification with real-world applications, this chapter provides readers with the essential groundwork to master advanced topics like toolpath optimization, macro programming, and 5-axis kinematics in subsequent chapters.

Chapter 2: CNC Programming

CNC FOCUS POINTS

This chapter explores the structure of CNC programs, essential G- and M-codes, programming formats, methods, programming cycles, best practices, verification methods, and advanced topics empowering engineers, machinists, and programmers alike. You will learn programming of turning, milling, turn-mill & 5axis machines. This book basically follows **Fanuc controller** with some references to **Siemens**.

2.1 Introduction

CNC machines operate through a standardized, machine-readable language composed of G-codes and M-codes. Originating in the 1950s at MIT and formalized internationally as ISO 6983, this language remains the foundation of CNC programming worldwide. Despite advances in CAM software automation, understanding G- and M-codes is essential to:

- Troubleshoot unexpected or erroneous machine behaviors.
- Optimize machining sequences to maximize efficiency.

- Write manual programs for prototyping or simple jobs.
- Edit CAM-generated code for custom needs.
- Interpret and program complex, multi-axis machining operations.

2.2 Fundamental Concepts

2.2.1 Categories of CNC Codes

- **G-Codes (Preparatory Codes):** Direct tool motions, coordinate system selections, and machining functions such as linear movements, arcs, and canned cycles.
- **M-Codes (Miscellaneous Codes):** Govern auxiliary machine functions including spindle control, coolant flow, and program control.

2.2.2 Program Structure Elements

- **Blocks:** Single lines of code acting as discrete instructions.
- **Modal Codes:** Stay active until altered (e.g., G01 for linear interpolation).
- **Non-Modal Codes:** Execute once per block (e.g., G04 for dwell).

2.2.3 Common Address Letters

- **X, Y, Z:** Linear axis coordinates.

- **A, B, C:** Rotary axis coordinates.
- **I, J, K:** Arc center offsets for circular interpolation.
- **F:** Feed rate (mm/min or in/min).
- **S:** Spindle speed (RPM).
- **T:** Tool number.
- **H:** Tool length offset number.
- **D:** Cutter radius compensation number.

2.3 Basic structure of a CNC Program

- **Header & Safety Block:** Part info, safety mode resets.
- **Tool Call & Spindle Start:** Tool selection, spindle spin-up, coolant activation.
- **Work Offset & Positioning:** Coordinate system selection, tool positioning.
- **Machining Operations:** Cutting movements (linear and circular).
- **Program Termination:** Tool retract, coolant and spindle shutoff, program end.

2.4 Fundamentals and Coordinate Systems

CNC programming transforms design concepts into precise machine instructions by defining specific points in three-dimensional space and choreographing the exact path the cutting tool takes between those points. This fundamental capability enables manufacturers to produce complex parts with tolerances measured in microns, with perfect repeatability across thousands of identical components.

2.5 Axes Definition

The Cartesian coordinate system forms the backbone of CNC programming. Understanding these axes is crucial for programming any CNC machine. For easier understanding, we will be discussing here a Vertical Machining Center (VMC).

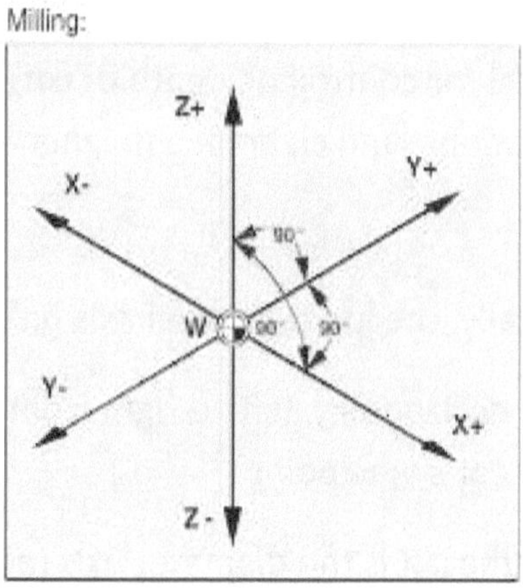

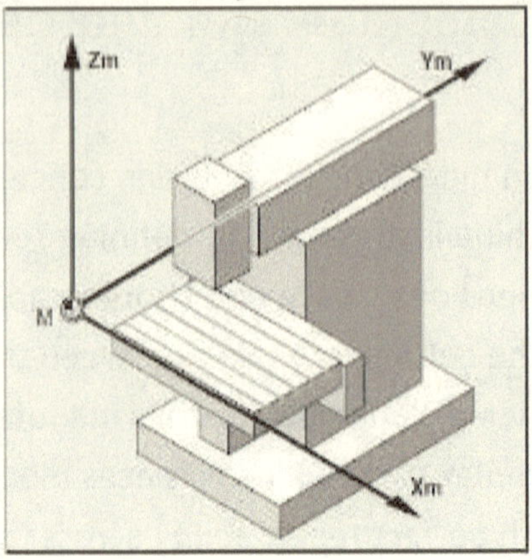

2.5.1 Z-Axis: The Spindle Axis

- Always aligned parallel to the spindle rotation axis
- Positive Z typically moves away from the workpiece (tool retraction)
- Negative Z moves toward or into the workpiece (cutting depth)
- Critical for controlling depth of cut, drilling operations, and clearance heights

2.5.2 X-Axis: The Longitudinal Axis

- Typically, the longest travel axis on the machine
- Runs horizontally, left-to-right from the operator's viewpoint
- On lathes, X is the diameter axis (perpendicular to spindle)

- Most commonly used for primary positioning moves

2.5.3 Y-Axis: The Transverse Axis

- Runs front-to-back in the horizontal plane
- Perpendicular to both X and Z axes
- On vertical machining centers, this moves the table toward/away from the operator
- Essential for 3-axis contouring and complex geometries

2.5.4 Additional Axes:

- **A, B, C axes:** Rotary axes around X, Y, and Z respectively
- **U, V, W axes:** Secondary linear axes parallel to X, Y, and Z
- Used in 4-axis, 5-axis, and multi-tasking machines

2.6 Quadrants and Directionality

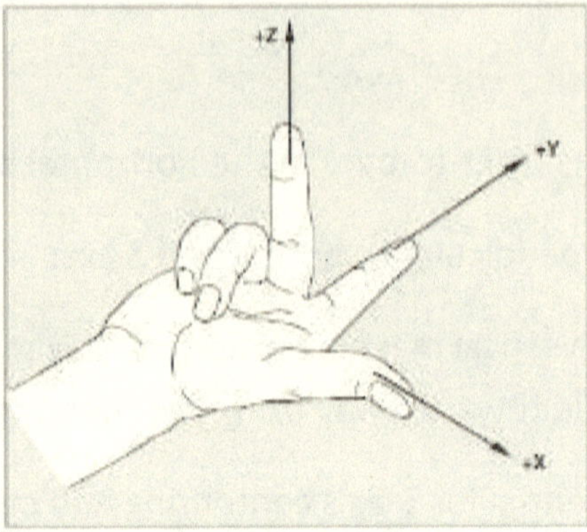

The **right-hand rule** provides an intuitive method for determining positive axis directions:

- Keep three fingers mutually perpendicular as shown in the figure
- Point middle finger towards the spindle along its axis (Z+)
- Your thumb will point towards X+
- Your forefinger will point towards Y+

  Important: While considering the positive direction of the axis, always consider the relative movement of the tool, irrespective of what physically moves. In some cases, the table will move and some cases tool / spindle will move.

This creates a standardized coordinate system with four quadrants in the XY plane:

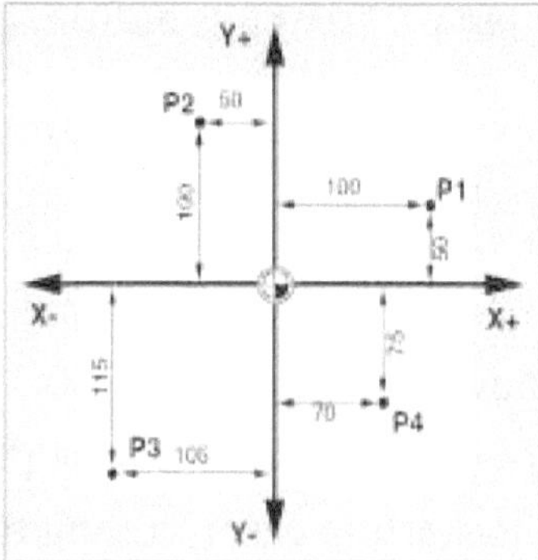

Quadrant 1 (Q1): X+, Y+

- Example coordinates (P1): X=100, Y=50
- Most common programming area

Quadrant 2 (Q2): X–, Y+

- Example coordinates (P2): X=–15, Y=+100
- Used when part extends left of origin
- Common in symmetrical part programming

Quadrant 3 (Q3): X–, Y–

- Example coordinates (P3): X=–105, Y=–115
- Used when part extends to left and bottom of origin
- Common in symmetrical part programming

Quadrant 4 (Q4): X+, Y–

- Example coordinates (P4): X=70, Y=–75

- Used when part extends bottom of origin
- Common in symmetrical part programming

Example: When machining a rectangular pocket centered at the origin, the tool path might begin at Q1 (X=50, Y=50), move through Q2 (X=–50, Y=50), continue to Q3 (X=–50, Y=–50), pass through Q4 (X=50, Y=–50), and return to the starting point, completing a counter-clockwise rectangular path.

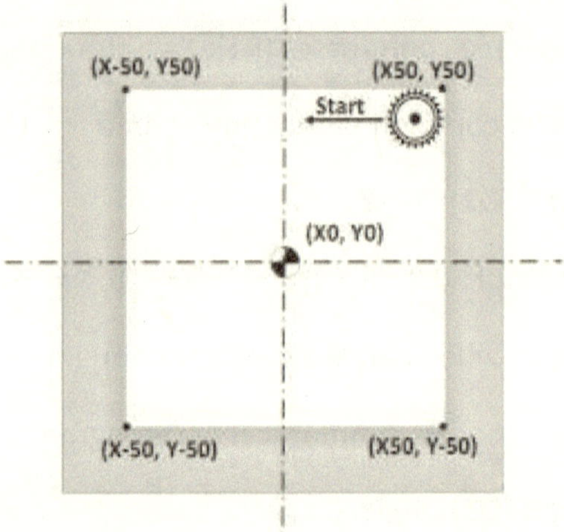

2.7 CNC Programming Language

CNC machines communicate through a standardized language consisting primarily of G-codes and M-codes, supplemented by coordinate values and parameters. Understanding this language is essential for effective CNC programming. It basically follows **ISO 6983**.

2.7.1 G-Codes (Preparatory Functions)

G-codes establish the operating mode of the machine, defining how subsequent commands will be interpreted. Many G-codes are modal, meaning they remain active until canceled by another code from the same group.

Common G-Code Groups:

Motion Modes (Group 01):

- G00 – Rapid positioning (non-cutting)
- G01 – Linear interpolation at feed rate
- G02 – Circular interpolation, clockwise
- G03 – Circular interpolation, counter-clockwise

Plane Selection (Group 02):

- G17 – XY plane (default for most mills)
- G18 – XZ plane (common for lathe work)
- G19 – YZ plane (specialized operations)

Work Coordinate Systems (Group 12):

- G54 through G59 – Work coordinate system 1 through 6
- Allows multiple part setups on one table
- Essential for tombstone and pallet work

Canned Cycles (Group 09):

- G81 – Drilling cycle
- G83 – Peck drilling cycle
- G73 – High-speed peck drilling
- Simplify repetitive operations

Feed Rate Modes (Group 05):

- G94 – Feed per minute (mm/min or in/min)
- G95 – Feed per revolution (mm/rev or in/rev)

Feed per minute = Feed per revolution x rpm

In milling machines, feed per minute / G94 is the default code where as in turning it is feed per revolution.

2.7.2 M-Codes (Miscellaneous Functions)

M-codes control auxiliary machine functions that don't involve coordinate movement.

Spindle Control:

- M03 – Spindle on, clockwise (CW) - standard for milling
- M04 – Spindle on, counter-clockwise (CCW) - left-hand tools/tapping
- M05 – Spindle stop

- M19 – Spindle orientation (specific angle positioning)

Coolant Control:

- M07 – Mist coolant on
- M08 – Flood coolant on
- M09 – All coolant off

Program Control:

- M00 – Program stop (requires operator intervention)
- M01 – Optional program stop (if enabled)
- M02 – Program end
- M30 – Program end and reset (returns to start)

Tool Management:

- M06 – Tool change
- M61 – Set tool number (without physical change)

Additional Functions:

- M98 – Call subprogram
- M99 – Return from subprogram
- M97 – Local subprogram call

2.8 Linear Movement Commands

Linear interpolation forms the foundation of CNC machining, enabling the machine to cut straight lines, create flat surfaces, and position tools accurately.

2.8.1 G00: Rapid Traverse

Function: Moves the tool at maximum machine speed without cutting **Characteristics:**

- Path may not be a straight line (depends on machine configuration)
- Each axis accelerates to its maximum rate independently
- Never use G00 when the tool is engaged in material
- Typical speeds: 10-30 m/min (400-1200 in/min)

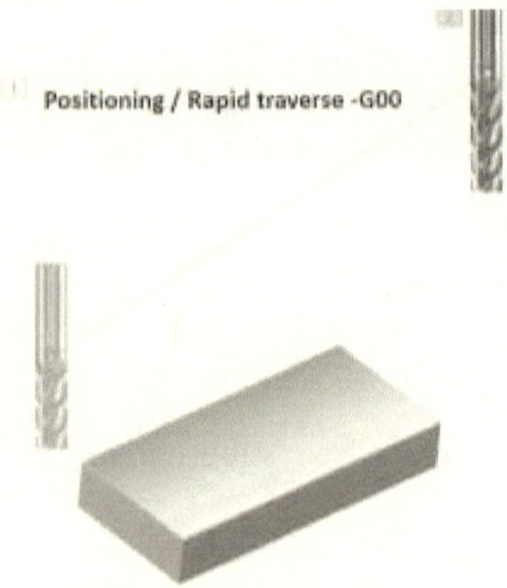

Syntax: G00 X___ Y___ Z___

Example:

G00 X100 Y50 Z50 (Rapid to clearance position)

G00 Z5 (Rapid to just above workpiece)

Safety Consideration: Always retract Z-axis before making large XY moves to avoid collisions with fixtures or workpiece.

2.8.2 G01: Linear Interpolation for cutting

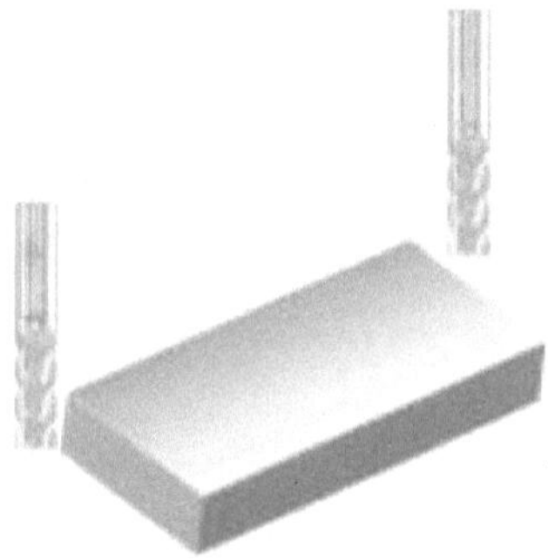

Function: Moves tool in a precisely controlled straight line at specified feed rate **Characteristics:**

- All axes move proportionally to reach endpoint simultaneously
- Path is always a straight line
- Feed rate (F) must be specified (in mm/min or in/min)
- Used for all cutting operations requiring straight edges

Syntax: G01 X___ Y___ Z___ F___

Movement Types:

Horizontal Movement (X-axis only):

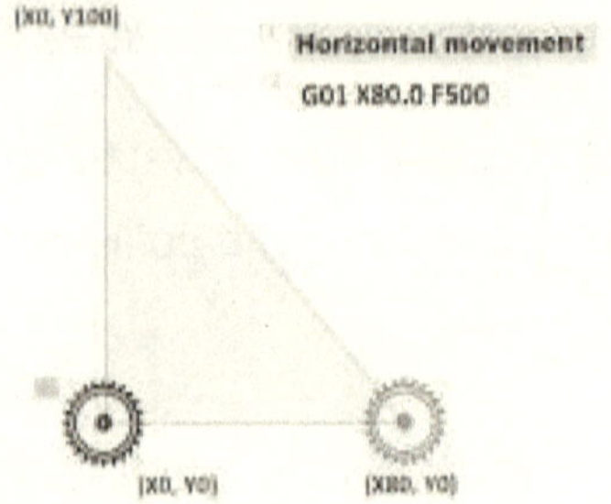

G01 X80.0 F500 (Move right to X=80 at 500mm/min)

Vertical Movement (Y-axis only):

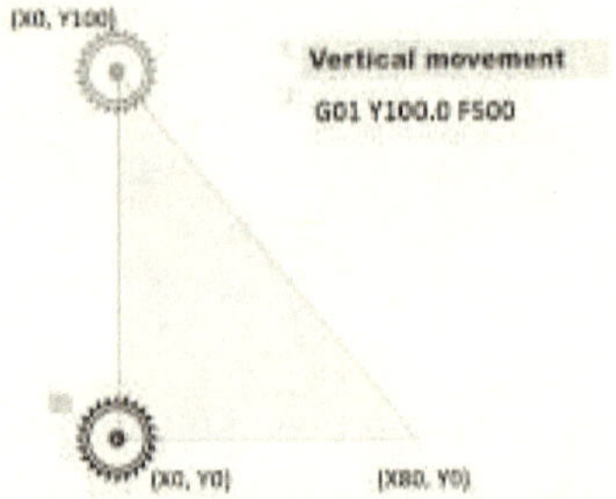

G01 Y100.0 F500 (Move to Y=100 at 500mm/min)

Angular Movement (simultaneous X and Y):

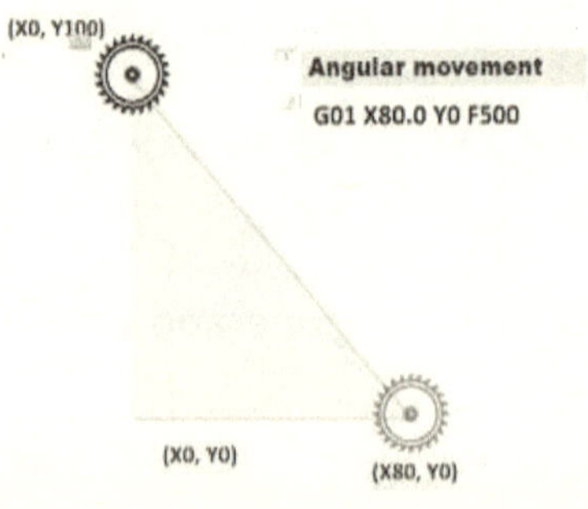

G01 X80.0 Y0 F500 (Move diagonally to X=80, Y=0)

3D Movement (X, Y, and Z simultaneously):

G01 X80. Y60. Z-10. F250. (Move to point in 3D space)

> **CAUTION**: A coordinate value without a decimal point may be treated as a value in microns. For example, X100 may be considered as X0.1

Practical Program Example:

Machining a rectangular pocket:

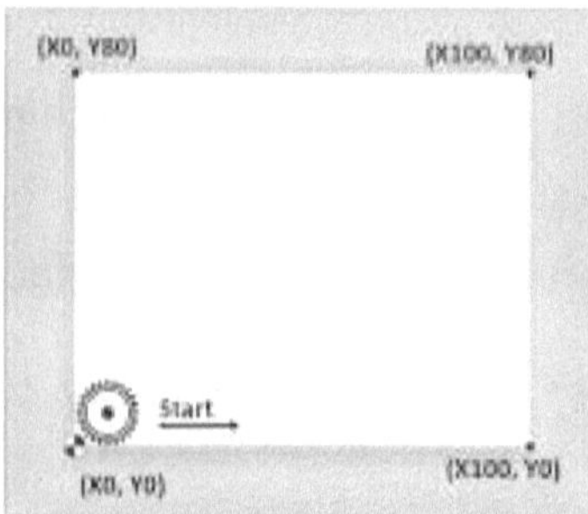

G00 X0 Y0 Z5. (Rapid to start position)

G01 Z-5. F100. (Plunge to depth)

G01 X100. F500. (Cut right edge)

G01 Y80. (Cut top edge)

G01 X0 (Cut left edge)

G01 Y0 (Cut bottom edge, complete rectangle)

G00 Z5. (Retract)

Feed Rate Considerations

The feed rate (F) value determines speed of cutting and affects:

- Surface finish quality (slower = finer finish)
- Tool life (excessive feed causes premature wear)
- Chip formation and evacuation
- Machine load and vibration

2.9. Circular Interpolation (G02 and G03)

Circular interpolation enables the machining of arcs and complete circles, essential for creating rounded features, chamfers, and contoured surfaces.

Understanding Circular Motion

2.9.1 G02: Clockwise Arc

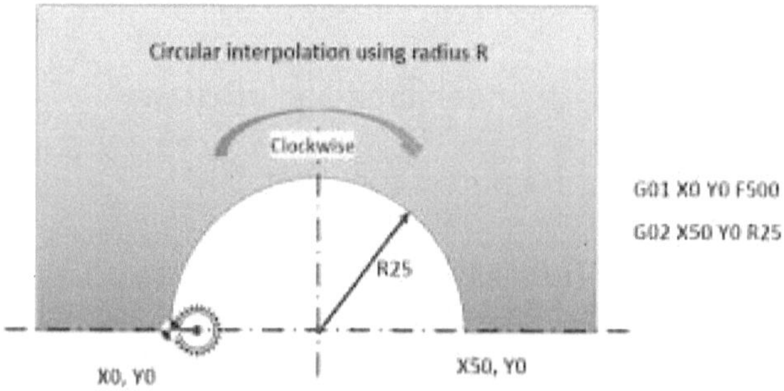

- Tool follows an arc in the clockwise direction
- Viewed from the positive side of the perpendicular axis

2.9.2 G03: Counter-Clockwise Arc

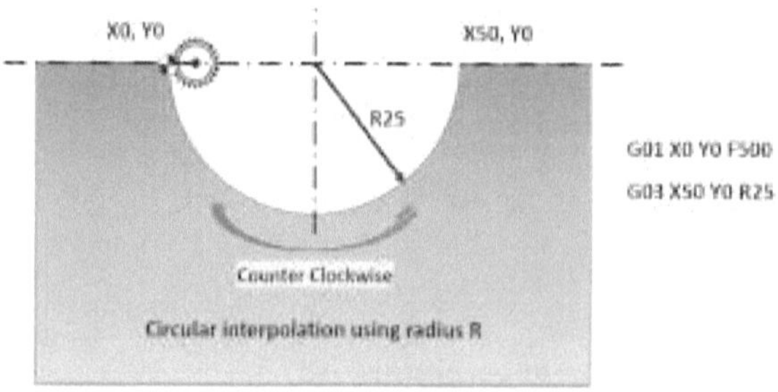

- Tool follows an arc counter-clockwise
- Mirror operation of G02

Important: Direction depends on your viewing angle. Always consider which plane (G17/G18/G19) is active.

Method 1: Radius Programming (R)

Syntax: G02/G03 X___ Y___ R___ F___

Advantages:

- Simpler to understand and program
- Fewer parameters required
- Good for arcs less than 180°

Limitations:

- Ambiguous for arcs greater than 180° (two possible solutions)
- Cannot program complete circles
- Less precise for CAM-generated code

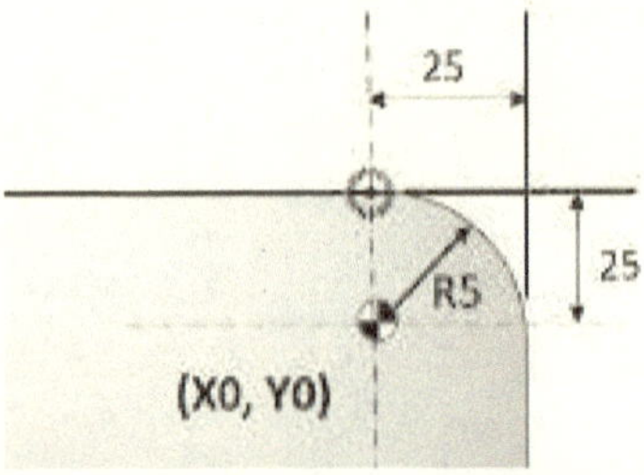

Example 1: Quarter-circle arc (90°)

G00 X0 Y25. (Start position)

G02 X25 Y0 R25. F300 (Clockwise 90° arc, radius 25mm)

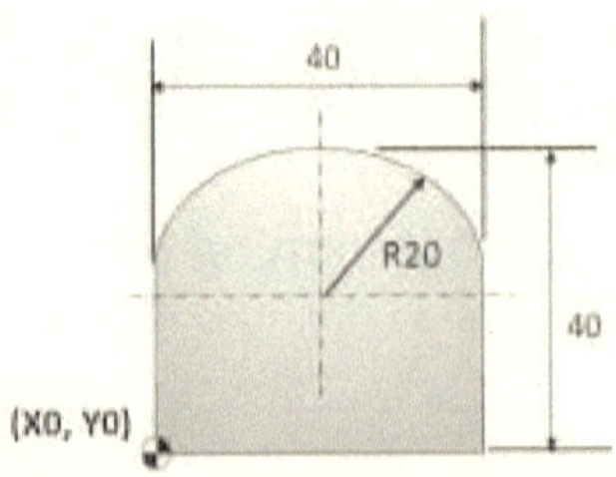

Example 2: Semi-circular slot

G01 X0 Y0 F500. (Move to start)

G01 Y20. (Straight up)

G02 X40. Y20. R20.

(180° arc - note: may cause errors on some controls)

G01 Y0 (Straight down)

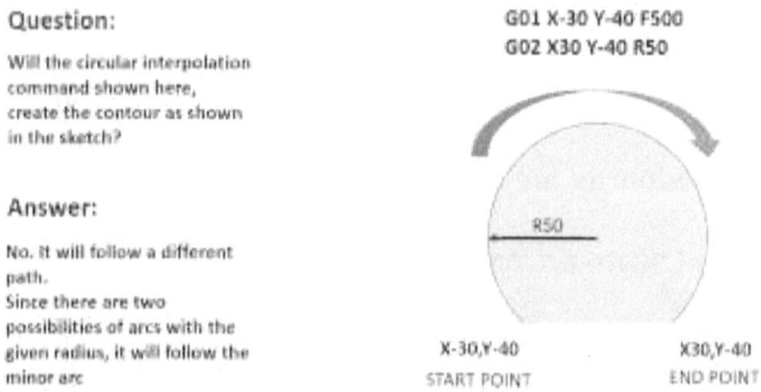

For arcs over 180°: Use negative radius value:

G02 X30. Y-40. R-50. F300. (Specifies the major arc)

Method 2: Center Point Programming (I, J, K)

Syntax: G02/G03 X___ Y___ I___ J___ K___ F___

Parameters:

- **I** = Incremental distance from start point to arc center in X

- **J** = Incremental distance from start point to arc center in Y
- **K** = Incremental distance from start point to arc center in Z

Advantages:

- Unambiguous for any arc length
- Can program full circles (start = end point)
- Preferred by CAM systems for precision
- Required for helical interpolation

Calculation Method:

1. Identify arc start point (current position)
2. Locate arc center point
3. Calculate: I = Center_X - Start_X
4. Calculate: J = Center_Y - Start_Y

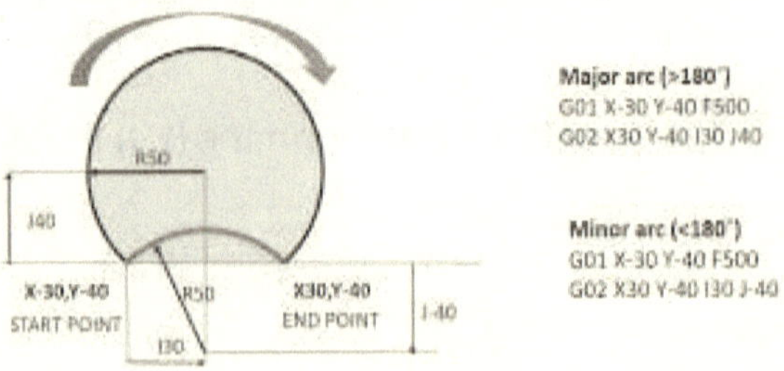

2.9.3 Programming arcs

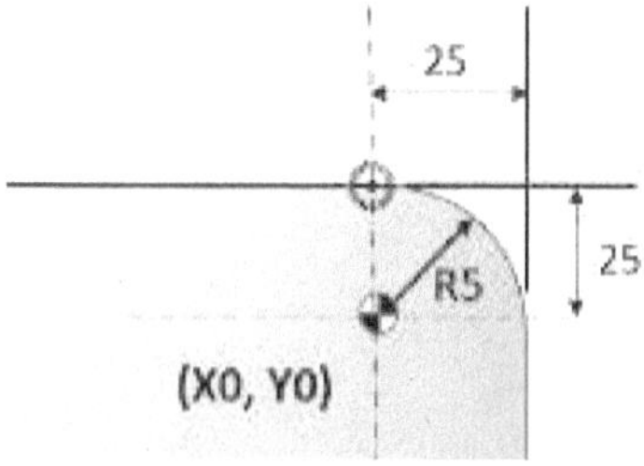

Example 1: Quarter-circle arc (90°)

G00 X0 Y25. (Start position)

G02 X25. Y0 I0 J-25. F300. (Clockwise 90° arc, radius 25mm)

Example 2: Semi-circular slot

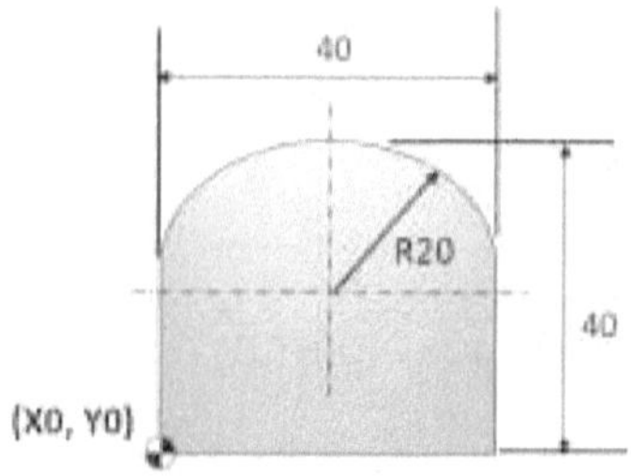

G01 X0 Y0 F500. (Move to start)

G01 Y20. (Straight up)

G02 X40. Y20. I20. J0

G01 Y0 (Straight down)

In the above example G02 command be also written as;

G02 X40. I20. (Since Y coordinate remain same and no distance in J)

2.9.4 Programming Full Circles

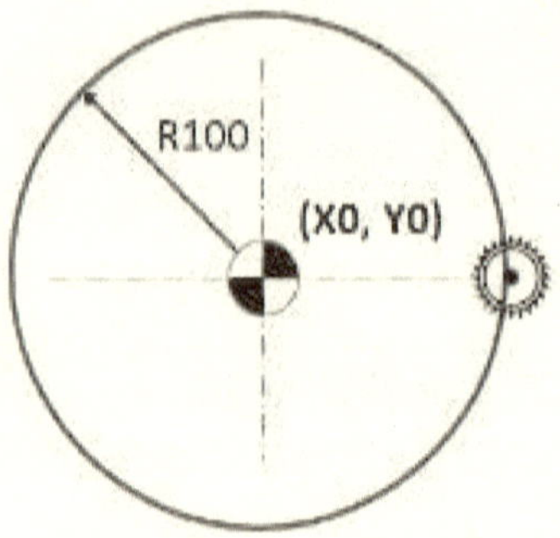

For complete circles, the start and end coordinates

are identical (often omitted).

Syntax: G02/G03 I___ J___ F___

Example 1: Circular boss (50mm diameter)

G00 X50. Y0 Z5. (Position at 3 o'clock position)

G01 Z-5. F100. (Plunge to depth)

G02 I-50. F600. (Complete circle, center at origin)

G00 Z5. (Retract)

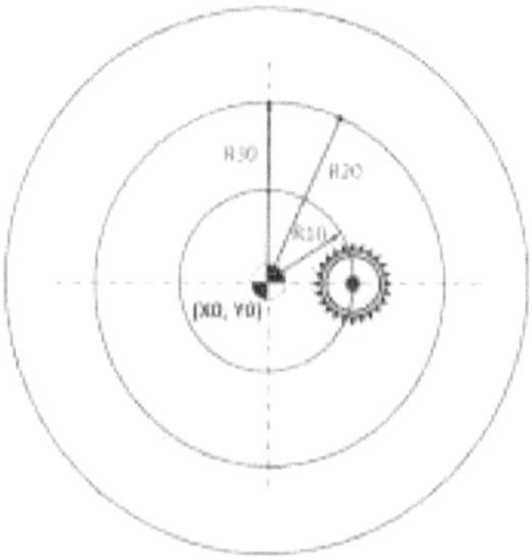

Example 2: Multiple circular passes for pocket

G00 X10. Y0 Z2. (Start at 10mm radius)

G01 Z-5. F80. (Plunge)

G02 I-10. F500. (Circle at 10mm radius)

G01 X20. (Move to 20mm radius)

G02 I-20. (Circle at 20mm radius)

G01 X30. (Move to 30mm radius)

G02 I-30. (Final circle at 30mm radius)

G00 Z5. (Retract)

2.9.5 Plane Selection Impact

The active plane (G17/G18/G19) determines which two axes form the circular motion:

- **G17 (XY plane):** Use I and J parameters
- **G18 (XZ plane):** Use I and K parameters
- **G19 (YZ plane):** Use J and K parameters

For more details on plane selection, refer section 2.14

2.10 Helical Interpolation

Helical interpolation represents an advanced programming technique that simultaneously combines circular motion in one plane with linear motion perpendicular to that plane, creating a three-dimensional spiral path.

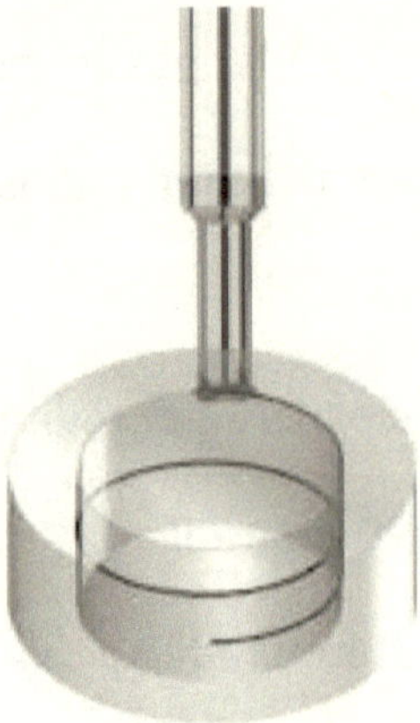

2.10.1 Understanding Helical Motion

Definition: Circular interpolation (G02/G03) with simultaneous movement in the perpendicular axis

Mathematical representation:

- 2D circular motion in one plane (e.g., XY)
- Linear motion in perpendicular axis (e.g., Z)
- Results in 3D helical path

Key Parameters:

- **Pitch:** The linear distance traveled per 360° revolution
- **Helix angle:** Angle of the spiral relative to the plane

2.10.2 Applications of Helical Interpolation

1. Thread Milling

- Internal and external threads
- Any diameter with one tool
- Left-hand or right-hand threads
- Superior to tapping for hard materials

2. Helical Ramping

- Gradual entry into pockets
- Reduces tool deflection
- Better chip evacuation
- Extends tool life

3. Specialized Features

- Spring forms and coils
- Helical grooves
- Spiral ramps
- Decorative patterns

2.10.3 Programming Helical Paths

Syntax: G02/G03 X___ Y___ Z___ I___ J___ F___

The Z parameter added to circular interpolation creates the helical motion.

Important: Combination of direction of circular movement and Z axis will define the hand of the thread.

G02 & Z- will create a right-hand thread and G03 Z- will create a left-hand thread

G03 & Z+ will create a right-hand thread and G02 Z+ will create a left-hand thread

Example 1: M12 × 1.75 Internal Thread Milling

(Thread: M12 x 1.75, depth 20mm)

(Tool: 10mm thread mill)

G00 X0 Y0 Z5. (Position to thread center)

G01 Z1. F100. (Move to thread start depth)

G01 G42 X6. Y0 (Position to thread start with cutter compensation)

(Helical thread cutting - 1.75mm pitch per revolution)

G02 X6. Y0 Z-0.75 I-6. J0 F150. (First revolution, descend 1.75mm; Z1 to Z-0.75)

G02 Z-2.5 I-6. (Second revolution. X & Y can be omitted as start and end are same)

G02 Z-4.25 I-6. (Third revolution)

G02 Z-6.0 I-6. (Fourth revolution)

(Continue for full depth...)

G02 Z-18.25 I-6. (Eleventh revolution)

G02 Z-20.0 I-6. (Final revolution)

G01 G40 X0 Y0 (Return to center)

G00 Z5. (Retract)

2.10.4 Thread Milling Advantages:

- One tool for multiple thread sizes
- No tool breakage concerns
- Can produce left-hand threads
- Better for interrupted threads
- Ideal for hard materials

Example 2: Helical Pocket Entry (2° Ramp Angle)

(Pocket: 50mm diameter, 15mm deep)

(Tool: 10mm end mill)

(Entry: Helical ramp at 2° angle)

(Calculate ramp parameters:)

(Circumference = π × 25mm = 78.54mm per revolution)

(Z per revolution at 2° = 78.54 × tan(2°) = 2.74mm)

G00 X25. Y0 Z2. (Position at pocket radius)

G01 Z0 F150. (Touch off surface)

(Helical entry ramp)

G03 X25. Y0 Z-2.74 I-25. J0 F400. (First revolution down 2.74mm)

G03 X25. Y0 Z-5.48 I-25. J0 (Second revolution)

G03 X25. Y0 Z-8.22 I-25. J0 (Third revolution)

G03 X25. Y0 Z-10.96 I-25. J0 (Fourth revolution)

G03 X25. Y0 Z-13.7 I-25. J0 (Fifth revolution)

G03 X25. Y0 Z-15.0 I-25. J0 F300. (Final partial revolution to full depth)

(Now at full depth, continue with pocket clearing)

G01 X23. F500. (Step in toward center)

G03 I-23. J0 (Clear at this radius)

(Continue pocket clearing pattern...)

Best Practices while using helical interpolation:

- Start with conservative feeds (70-80% of normal)
- Ensure adequate coolant for chip evacuation

- Use climb milling direction when possible
- Verify helix angle doesn't exceed tool capabilities
- Check machine capability (not all controls support helical interpolation)

Key takeaways:

- Begin programs with safety blocks to reset modal states.
- Understand modal vs. non-modal codes for predictable machine actions.
- Structure programs with clear headers, ample commenting, and safe termination.
- Verify code thoroughly via simulation, dry runs, and single-step execution.
- Remain updated on control-specific features and evolving CNC programming practices

2.11 G04 – Dwell Time Command

The **G04** command is used to **pause machine motion for a specified period**, allowing time for certain operations to complete before the program continues. This command is often referred to as a *dwell*.

Purpose

A dwell is typically used to:

- Allow the spindle to stabilize at the commanded speed before cutting begins.
- Ensure a tool stays in contact with the workpiece for a fixed time (e.g., during drilling or tapping).
- Provide time for coolant or chip evacuation between cycles.
- Allow a probe or measuring device to settle before taking a reading.

Format

G04 P___ ; or G04 X___ ;

- **P** specifies the dwell time in **milliseconds** on most Fanuc-type controls (e.g., P1000 = 1 second).
- **X** can also specify dwell time in **seconds** depending on the machine configuration.

Example 1:

G04 P2000 ;

The program pauses for **2 seconds** before proceeding.

Example 2 (Drilling Dwell):

G01 Z-10. F100 ;

G04 P1000 ;

G0 Z10. ;

The tool dwells for 1 second at the bottom of the drilled hole.

Notes

- Dwell is executed **after** the previous motion command is completed.
- Excessive dwell time can cause **tool wear** or **surface burnishing**.
- Always verify the control's dwell unit (seconds vs milliseconds) in the machine parameter settings.

2.12 Absolute vs. Incremental Programming

2.12.1 Absolute Positioning (G90)

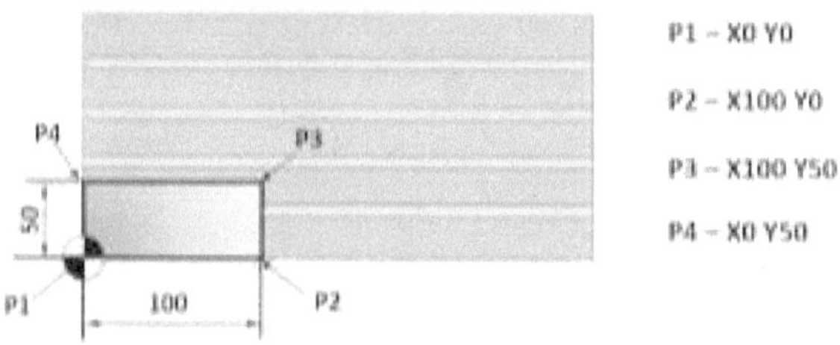

- All coordinates reference the work coordinate system origin
- Most common and recommended for clarity

2.12.2 Incremental Positioning (G91)

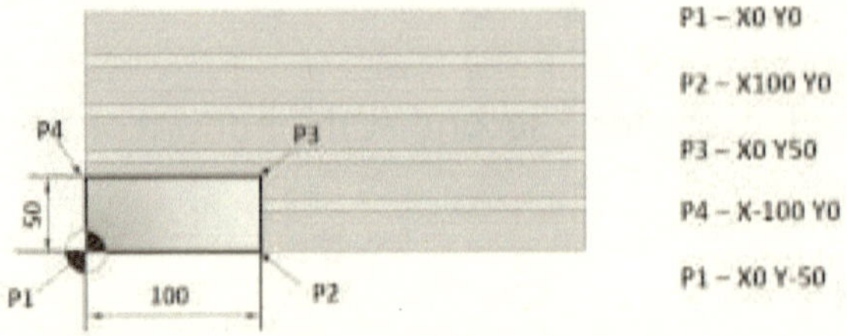

- Coordinates reference the current tool position. Let us analyze P4. The X distance from P3 to P4 is 100 and it is in minus direction. The Y distance is zero. For same point P4, absolute coordinate is X0, Y50
- Useful for repetitive patterns and sub-routines (refer section 2.15 for more details)

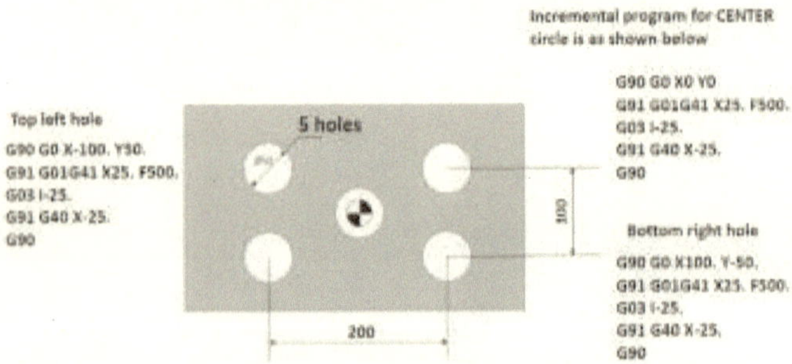

In the above example, you can notice that basic program for all the holes is same. This is the benefit of using an incremental command. This can be further simplified by using a subprogram. Please refer the section 2.15 on subprogram for more details

2.13 Inch and Metric System

CNC machines worldwide operate using two primary systems of measurement: **Inch (Imperial)** and **Metric (SI)**. Understanding these systems and how to switch between them is essential for ensuring dimensional accuracy, tool selection, and compatibility with customer drawings.

2.13.1 Overview of Units

- **Inch System (Imperial):**
  Used primarily in the United States and certain industries like aerospace and defense. Dimensions are expressed in inches, and feeds/speeds are often defined in **inches per minute (IPM)**.

- **Metric System (SI):**
  Used internationally and in most modern manufacturing environments. Dimensions are expressed in millimeters (mm), with feeds and speeds typically in **millimeters per minute (mm/min)**.

2.13.2 CNC Codes for Unit Selection

Most CNC controls, including Fanuc, use standard G-codes to define the active measurement system:

- **G20 – Inch Input Mode**
  Activates the inch measurement system. All

coordinate and feed rate values are interpreted as inches.

- **G21 – Metric Input Mode**
  Activates the metric measurement system. All coordinate and feed rate values are interpreted as millimeters.

Example:

G20 (Set inch mode)

G00 X1.0 Y2.0

G01 Z-0.25 F5.0

Here, all values are in inches.

G21 (Set metric mode)

G00 X25.4 Y50.8

G01 Z-6.35 F125.0

Here, the same motion is programmed in millimeters.

Note: Some controllers like Siemens follow G70 for Imperial and G71 for Metric.

2.13.3 Importance of Unit Consistency

Switching between inch and metric without proper conversion can result in dimensional errors or machine crashes. Always verify:

- The **drawing unit system** (metric or inch).

- The **machine's default unit setting** (check parameter or control screen).
- Tool length and diameter values are in the same unit system.
- Feed rates, spindle speeds, and offsets are converted appropriately.

2.13.4 Converting Between Inch and Metric

Use the following conversions when required:

- **1 inch = 25.4 millimeters**
- **1 millimeter = 0.03937 inches**

Example:
A toolpath programmed in inches for X2.0 must be converted to X50.8 when using metric mode.

2.13.5 Best Practices

- Standardize all programs and setups to **one measurement system** in a given job.
- Include a **G20 or G21** command at the start of every program to eliminate ambiguity.
- Verify the **CAM post-processor** output matches your preferred unit system.
- When importing CAD/CAM models, confirm unit scaling to prevent machining errors.

2.14 CNC Plane Selection

2.14.1 Introduction

Plane selection is a fundamental concept in CNC programming that provides a strategic instruction to the machine's controller. Its primary importance lies in ensuring the correct interpretation of geometric commands, which is especially critical for the unambiguous execution of circular interpolation (G02/G03) and cutter radius compensation (G41/G42). The purpose of G-code plane selection commands is to definitively inform the controller about the active two-dimensional working plane, which in turn dictates the axes used for these critical operations.

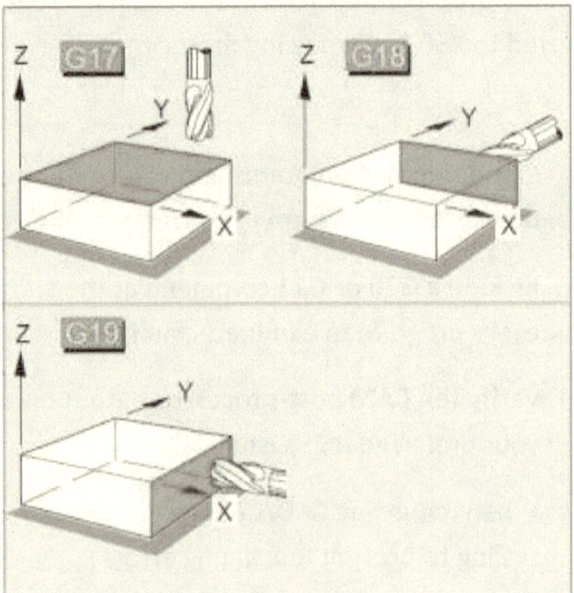

The three primary G-codes used for this purpose are:

- **G17:** Selects the X-Y Cartesian plane.
- **G18:** Selects the X-Z Cartesian plane.
- **G19:** Selects the Y-Z Cartesian plane.

While these commands are foundational, they are not required for every operation. The following sections will detail the specific conditions under which these commands become a critical component of a robust CNC program.

2.14.2 Necessity of Plane Selection

While a plane is always active by default, explicitly defining it within a program is not always necessary. A programmer must understand precisely when a plane selection command is mandatory to prevent geometric errors and ensure the controller behaves as intended. Explicit plane selection is mandatory for any operation where the controller must resolve a two-axis motion, such as defining the center of an arc or the offset direction for tool compensation.

The following operations **mandate** the use of a plane selection command:

- **Circular Interpolation:** To define the plane in which an arc (G02/G03) is machined and to determine which two of the three vector addresses (I, J, K) are used to define the arc's center point.

- **Drilling Cycles:** To define the two-dimensional plane (e.g., X-Y) where hole locations are programmed, making the third axis (e.g., Z) the unambiguous axis for the drilling and retraction motions of the cycle.
- **Radius Compensation:** To correctly apply cutter radius compensation (G41/G42) relative to the toolpath in the defined two-axis plane.

Conversely, for operations such as plain, straight-line milling, an explicit plane selection command is not required. The machine can perform these movements irrespective of the active G17, G18, or G19 command. Understanding this distinction is key to writing efficient and error-free code.

2.14.3 Plane Selection for Milling

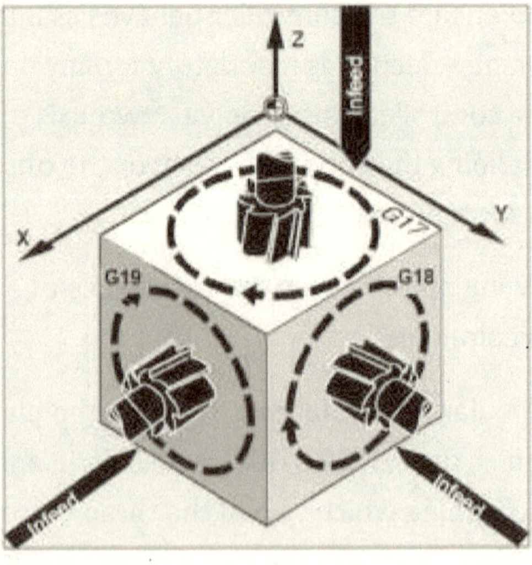

On a Vertical Machining Center (VMC), understanding the default operational plane and the relationship between the working plane and its corresponding in-feed axis is crucial for both standard and advanced milling. Upon switch-on, the default active plane for a VMC or standard milling machine is the **G17 (X-Y) plane**.

The relationship between each G-code, its working plane, and the corresponding in-feed axis on a VMC is as follows:

G17 (X-Y Plane) This is the most common working plane for a VMC. When G17 is active, circular and compensation moves occur on the X-Y plane, and the **Z-axis** serves as the perpendicular axis for tool depth control, peck drilling cycles, and Z-level finishing passes.

G18 (X-Z Plane) When the G18 plane is selected, the working plane is defined by the X and Z axes. Consequently, the **Y-axis** becomes the perpendicular in-feed axis, used for operations like side-milling arc profiles with a right-angle head.

G19 (Y-Z Plane) When the G19 plane is selected, the working plane is defined by the Y and Z axes. In this configuration, the **X-axis** becomes the perpendicular in-feed axis.

While G17 is the standard for most VMC work, a programmer must call G18 or G19 for specialized applications. A common scenario requiring a non-default plane is when using a **right-angle head**. This attachment changes the cutter's orientation, requiring milling operations where the X or Y axis must act as the in-feed direction.

2.14.4 Plane Selection for Turning

Turning centers operate with a different default axis configuration than milling machines. Understanding this default is essential for standard turning as well as for advanced turn-mill operations. On a standard turning center, the **Z-axis** is coaxial with the main spindle, and the **X-axis** controls the diameter.

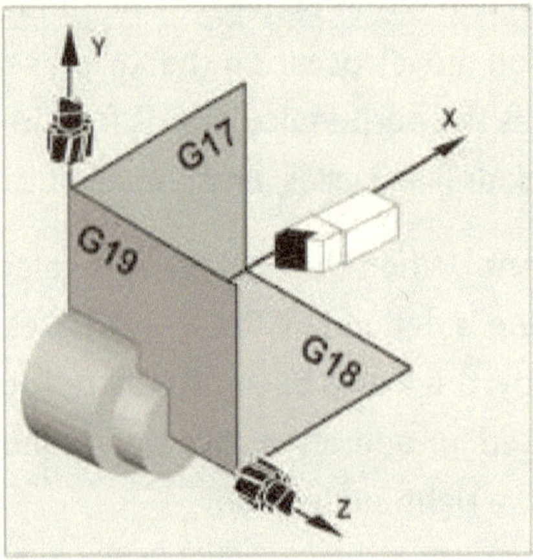

The default active plane for a turning center is the **G18 (X-Z) plane**. This default accommodates the vast majority of standard turning, grooving, and facing operations where circular interpolation and radius compensation occur between the primary X and Z axes.

Scenarios requiring non-default plane selection (G17 or G19) arise when the turning center is equipped with live tooling and is capable of milling. For example, selecting **G17 (X-Y plane)** is necessary for milling a bolt-hole pattern on the face of the part, while **G19 (Y-Z plane)** would be used for milling a keyway along the part's outer diameter using a live cross-working tool. In these cases, the programmer must explicitly select the appropriate plane to ensure the controller correctly interprets the milling toolpaths.

2.14.5 Quick Reference Guide

The following table synthesizes the key information from this manual into a single, easy-to-use reference. It provides a quick overview of each plane selection command, its associated axes, and its default machine application for programmers and machinists.

G-Code	Working Plane	Associated In-Feed Axis	Default Machine Type
G17	X-Y	Z-axis	VMC / Milling Machine
G18	X-Z	Y-axis (on VMC)	Turning Center
G19	Y-Z	X-axis	Not specified as a default

2.15 Subprogram / Subroutine

Subprograms in CNC programming are powerful tools that allow a programmer to **reuse blocks of code**, simplify complex programs, and reduce human error. Instead of rewriting repetitive machining cycles multiple times, a subprogram can be written once and called whenever required.

This concept is similar to using “functions” or “subroutines” in conventional programming languages. Subprograms make CNC programs more structured, easier to debug, and faster to modify.

A **subprogram** is a secondary program that performs a specific, reusable operation—such as drilling a hole pattern, machining a slot, or profiling a feature—and can be called by a **main program** (also called the parent or master program).

When the CNC control encounters a subprogram call command, it **temporarily transfers control** from the main program to the subprogram. After completing the subprogram, the control **returns automatically** to the main program at the point just after the subprogram call.

2.15.1 Key G-Codes and Commands

Function	**G/M Code**	**Description**
Subprogram call	M98 Pxxxx	Calls subprogram number Oxxxx

Function	G/M Code	Description
Subprogram end and return	M99	Ends subprogram and returns to the main program
Looping a subprogram	M98 Pxxxx Ln	Calls subprogram Oxxxx and repeats it n times
Program end	M30	Ends the main program and resets the control

2.15.2 Structure of a Subprogram

A subprogram is written just like a normal CNC program but has its **own program number** (e.g., O1000, O2000, etc.).
It must **end with M99**, not M30, because M99 tells the control to return to the calling program.

2.15.3 Example1: Face milling

O1000 (Main Program FOR FACE MILLING)

...

G90 G54 G0 X0 Y0

M98 P2000 L3 (Call subprogram O2000 three times)

...

M30

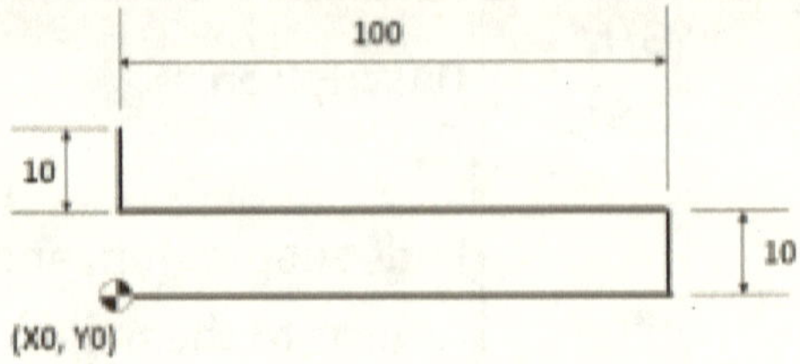

O2000 (Subprogram FOR FACE MILLING)

G90

G01 X100.0 F500 (Machining along X axis)

G91 Y10.0 (Incremental movement in Y axis)

G90 X0.0 (Machining along X axis)

G91 Y10.0 0 (Incremental movement in Y axis)

G90

M99

Explanation: This subprogram will do face milling for two step over with total 20mm removal in Y direction. When subprogram call it 3 times, it will do milling for 60mm.

2.15.4 Example2: Drilling patterns on Multiple Locations

Suppose you need to drill the **same hole pattern at three different locations** on a plate.

Main Program – O0100

O0100

```
G90 G17 G54

T01 M06

S1500 M03

G43 H01 Z100.

G00 Y0.

M98 P0200 (Call hole pattern subprogram)

G00 Y50.

M98 P0200

G00 Y100.

M98 P0200

G00 Z100.

M30
```

Subprogram – O0200

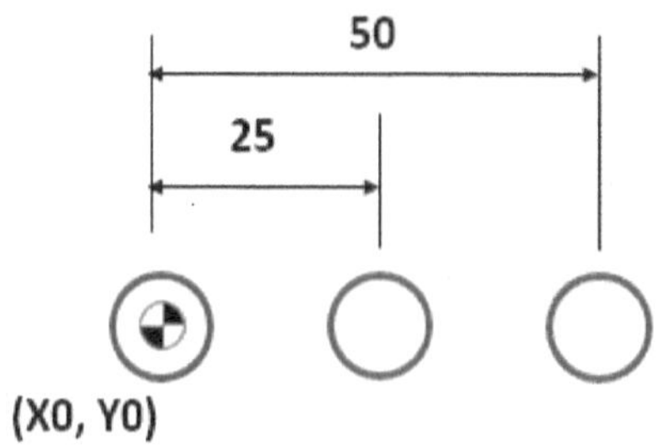

```
O0200 (Hole Pattern)

G90 G0 X0.

G81 R2. Z-10. F120.

G80

G00 X25.

G81 R2. Z-10. F120.

G80

G00 X50.

G81 R2. Z-10. F120.

G80

M99
```

Explanation:

- The subprogram O0200 defines a 3-hole pattern (0, 25, 50 mm apart).
- The main program calls this subprogram at Y=0, Y=50, and Y=100 positions—creating three identical hole patterns.

2.15.5 Example3: Interpolation of holes on multiple Locations

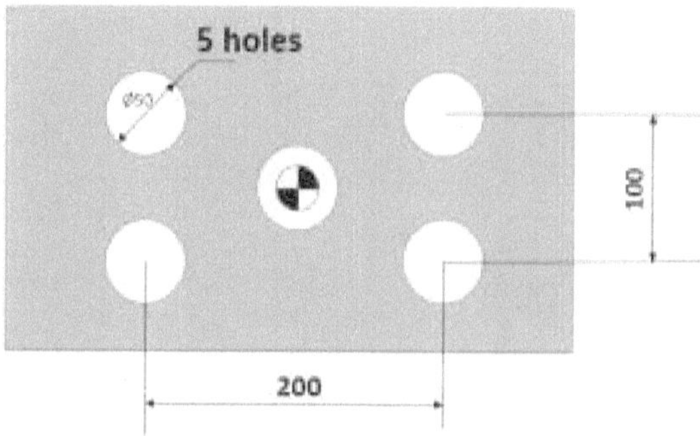

Main Program – O5000

O0100

G90 G17 G54

T01 M06

S4000 M03

G43 H01 Z100.

G00 X0 Y0.

M98 P0500 (Call interpolation subprogram)

G00 X-100. Y50.

M98 P0500

G00 X-100. Y-50.

M98 P0500

G00 X100. Y-50.

```
M98 P0500

G00 X100. Y50.

M98 P0500

G00 Z100.

M30
```

Subprogram – O0500

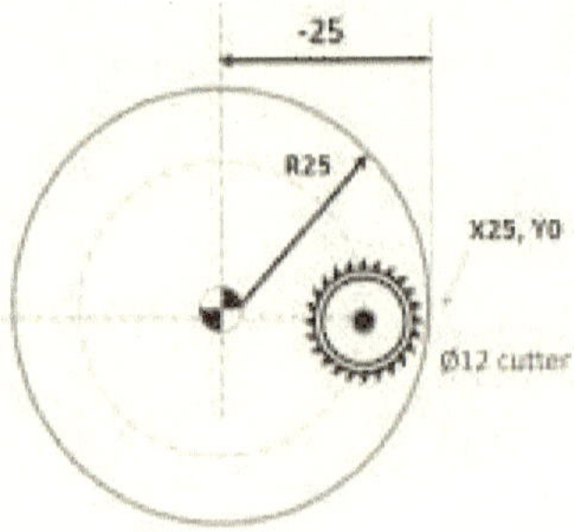


```
O0500 (Hole interpolation)

G90 G0 Z2.

G01 Z-5. F100.

G91 G01 G41 X25. F500.

G03 I-25.

G91 G40 X-25.

G90 G0 Z100.

M99
```

2.15.6 Subprograms Within Subprograms (Nested Subprograms)

Subprograms can **call other subprograms**, known as *nesting*. However, excessive nesting can make debugging complex and should be used carefully.

Example:

```
O1000 (Main)

M98 P2000

M30

O2000

M98 P3000

M99

O3000

...

M99
```

2.15.7 Advantages of Using Subprograms

- **Reusability:** Write once, use multiple times.
- **Compact code:** Reduces program size and memory usage.
- **Consistency:** Ensures identical machining features across parts.
- **Ease of modification:** Changing one subprogram updates all instances automatically.
- **Better readability:** The main program becomes cleaner and more structured.

2.15.8 Notes and Best Practices

- Always **use meaningful comments** for clarity.
- Avoid **too many nested levels** (most controls allow up to 4–10).
- Ensure **unique program numbers** to prevent accidental overwrites.
- Verify subprograms independently before linking to main programs.
- In Fanuc controls, subprogram numbers usually start from O0001 to O9999.

2.15.9 Key Takeaways

Subprograms are an elegant solution for repetitive machining operations. They make programs modular, reduce duplication, and minimize error. Once mastered, subprogram techniques are indispensable for professional CNC programmers—especially when machining complex parts with multiple repeated patterns or identical features.

2.15.10 Professional Tips for CNC Programmers

Before Running Any Program:

1. **Simulation First**: Always run programs through CAM software simulation or the machine's dry-run/graphics mode before cutting. A single sign error (+ vs –) can result in costly crashes.

2. **Tool Path Verification**: Check for:
    - Rapid moves that might collide with fixtures
    - Adequate clearance heights
    - Correct work coordinate system selection
    - Tool length offsets applied correctly
3. **Feed and Speed Validation**: Verify that all F (feed) and S (speed) values are appropriate for:
    - Material being machined
    - Tool diameter and type
    - Depth of cut
    - Machine capabilities
4. **Safety Checks**:
    - Confirm all M-codes are supported by your machine
    - Verify spindle direction (M03 vs M04)
    - Check coolant availability
    - Ensure emergency stop is accessible

2.15.11 Programming Best Practices:

Use Comments Liberally:

(PROGRAM: BRACKET_V2.3)

(MATERIAL: 6061-T6 ALUMINUM)

(SETUP: VISE, PARALLEL TO X-AXIS)

(DATUM: TOP CENTER OF PART)

N10 (TOOL 1: 12MM END MILL, ROUGHING)

Organize Code Logically:

- Group similar operations
- Use sequence numbers (N10, N20, etc.)
- Separate tool changes with blank lines
- Include tool descriptions

Implement Safety Zones:

G28 G91 Z0 (Return Z to machine home safely)

G90 (Return to absolute positioning)

Test New Programs Carefully:

- Run first part at 50% feed rate override
- Keep hand near feed rate and rapid override controls
- Watch first few operations closely
- Verify dimensions on first part before production run

2.15.12 Common Programming Errors to Avoid:

1. **Forgetting plane selection** before circular interpolation
2. **Mixing absolute and incremental** mode inadvertently
3. **Omitting feed rate** on first G01 after G00
4. **Incorrect I, J, K calculation** causing spiral paths
5. **Rapid positioning (G00) while tool is in material**
6. **Missing tool length compensation** activation
7. **Wrong work coordinate system** (G54 vs G55, etc.)

2.15.13 Machine-Specific Considerations:

Always consult your specific machine manual, as:

- G-code definitions can vary between manufacturers
- Some machines use different code formats (e.g., FANUC vs. Siemens)
- M-code numbers may differ for the same function
- Parameter addressing can be unique
- Canned cycles may have different syntax

2.16 Cutter Radius Compensation (CRC)

Introduction: What is Cutter Radius Compensation?

Cutter Radius Compensation (CRC) is one of the most powerful and essential features in CNC programming. It allows the programmer to define the exact part geometry without worrying about the radius of the cutting tool. Once CRC is activated, the CNC control automatically shifts the programmed tool path to account for the tool's radius.

In practical terms, CRC ensures that it is always the cutting edge of the tool—not its center—that follows the programmed contour. This capability saves the programmer from tedious offset calculations, dramatically improves accuracy, and provides flexibility when changing to tools of different diameters without modifying the part program.

2.16.1 The Core Problem: Tool Center vs. Part Edge

CNC machines naturally follow the path of the tool center point. However, when machining, it's the cutting edge of the tool that defines the material left behind and determines the final part dimensions.

Example Scenario: The Undersized Circle Problem

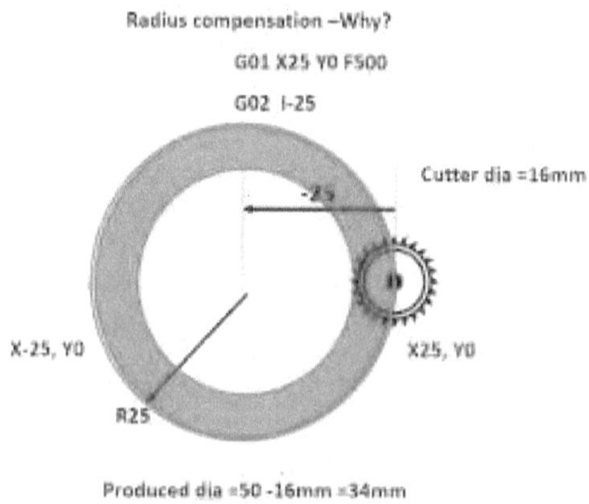

Consider this common machining task:

- **Desired part profile:** A circular boss of Ø50 mm (radius = 25 mm)
- **Tool used:** Ø16 mm end mill (radius = 8 mm)

If the tool's center is commanded to follow the 25 mm radius path, the cutting edge will intrude into the part profile by the tool radius. The resulting part will actually measure:

Final Diameter = Programmed Diameter – 2 × Tool Radius = 50 – 16 = 34 mm

Instead of a precise 50 mm circular boss, you end up with a severely undersized 34 mm diameter—a 16 mm dimensional error that renders the part scrap.

Programming Approach	Undesired Result
Tool center follows the intended 50 mm circle	Part cut to Ø34 mm instead of Ø50 mm
Material removal: 8 mm too deep on all sides	Part rejected by quality control

2.16.2 The Manual Compensation Nightmare

Without CRC, programmers would need to manually calculate offset paths:

- For external profiles (islands): Offset path outward by tool radius
- For internal profiles (pockets): Offset path inward by tool radius
- For complex contours with arcs and tangencies: Calculate new arc centers, intersection points, and tangent transitions

This manual approach is:

- **Time-consuming:** Hours of trigonometric calculations
- **Error-prone:** Easy to make mistakes in sign or magnitude
- **Inflexible:** Changing tool diameter requires complete reprogramming

- **Maintenance nightmare:** Tool wear compensation becomes impossible

This fundamental mismatch between tool center motion and desired part geometry is exactly the problem CRC is designed to solve.

2.16.3 The Solution: How Compensation Works

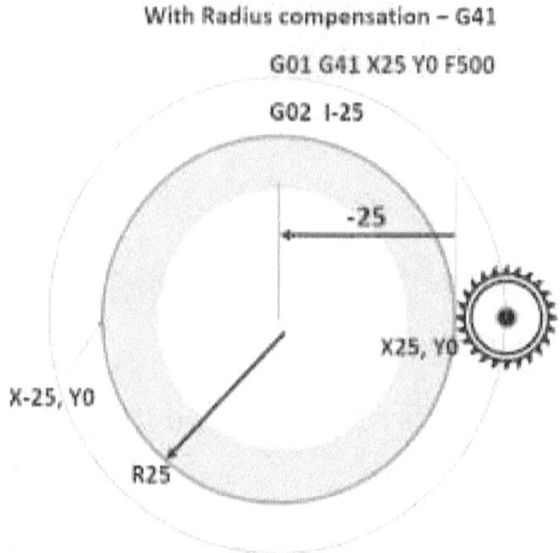

With CRC enabled, the CNC control automatically offsets the tool path outward (or inward) by the tool radius stored in the tool offset table. The programmer can then:

1. **Focus on programming the true part geometry** – Code the profile exactly as dimensioned on the drawing
2. **Avoid manual trigonometry** – Eliminate error-prone radius adjustments and offset calculations

3. **Swap tools flexibly** – Change to different diameter tools without rewriting the program (only the tool offset table needs updating)
4. **Compensate for tool wear** – Adjust the tool radius value in the offset table to maintain part accuracy

This approach not only simplifies programming but also enhances flexibility during production, tool changes, and rework operations.

2.16.4 How the Control Calculates the Offset

The CNC control performs sophisticated lookahead calculations:

1. **Reads the current block** and the next block(s) ahead
2. **Determines the offset direction** based on G41/G42 and motion direction
3. **Calculates intersection points** where offset paths meet
4. **Generates the actual tool path** perpendicular to the programmed path at the tool radius distance

This happens in real-time, creating smooth transitions even at sharp corners and tangent arcs.

2.16.5 Putting Compensation into Practice: Key G-Codes

Cutter compensation is activated, directed, or canceled using three essential G-codes: G41, G42, and G40.

2.16.5.1 G41 & G42 – Left and Right Compensation

G41: Compensation Left – Offsets the tool to the left of the programmed path (relative to direction of motion).

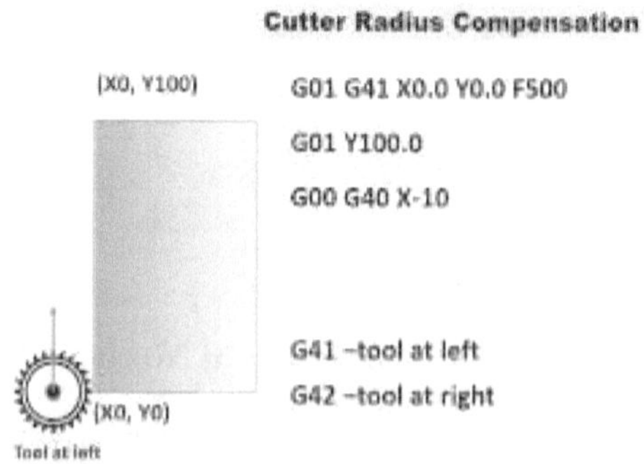

G42: Compensation Right – Offsets the tool to the right of the programmed path (relative to direction of motion).

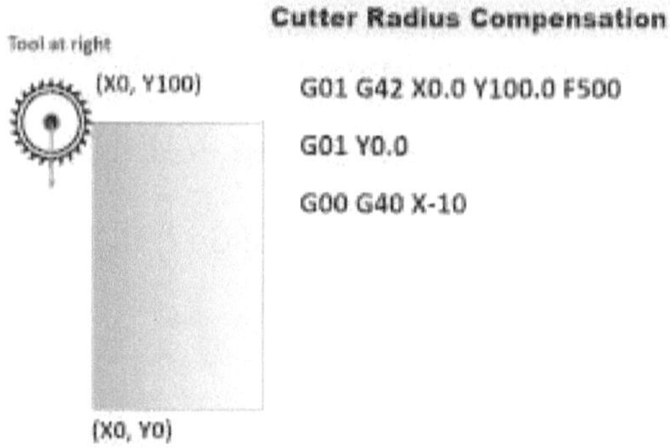

Rule of Thumb: Imagine yourself "walking along" the programmed path in the tool's travel direction:

- **G41** = tool positioned on your left
- **G42** = tool positioned on your right

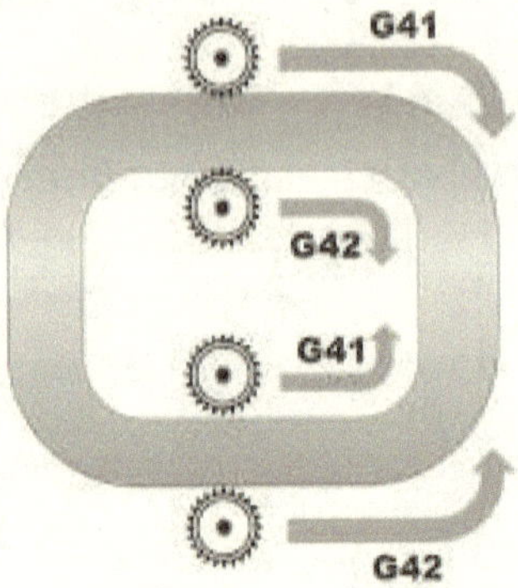

Determining G41 vs. G42: A Visual Method

When looking down at the workpiece (Z-axis view):

For External Profiles (cutting around an island):

- Traveling counterclockwise → Use G42
- Traveling clockwise → Use G41

For Internal Profiles (cutting inside a pocket):

- Traveling counterclockwise → Use G41
- Traveling clockwise → Use G42

2.16.5.2 G40 – Canceling Compensation

At the end of the contoured section, you must cancel compensation with **G40**. This returns the control to normal (uncompensated) mode.

Critical Warning: Forgetting to cancel compensation can lead to unexpected offsets in subsequent moves—sometimes resulting in collisions, scrapped parts, or machine damage.

2.16.6 How Compensation Looks in a CNC Program

A CRC code is not used in isolation—it is part of a motion command within a program block.

Typical Block Structure

N100 G41 G01 X50.0 Y25.0 F150.0

Block elements explained:

- **N100** – Block number (sequence number for reference)
- **G41** – Activate compensation left
- **G01** – Linear interpolation move
- **X50.0 Y25.0** – Target coordinate position
- **F150.0** – Feed rate in mm/min

The offset will remain active for all subsequent blocks until explicitly canceled with G40.

2.16.7 Complete Programming Example: Rectangular Boss

Let's machine a rectangular Boss with the following specifications:

Part Requirements:

- Boss dimensions: 90 mm × 50 mm
- Corner radius: 5 mm
- Mill depth: 10 mm below surface
- Material: Aluminum 6061

Tooling:

- Tool: Ø12 mm carbide end mill (radius = 6 mm)
- Tool number: T03
- Stored radius in offset table: 6.0 mm

2.16.8 Program Without CRC (Manual Offset – Tedious):

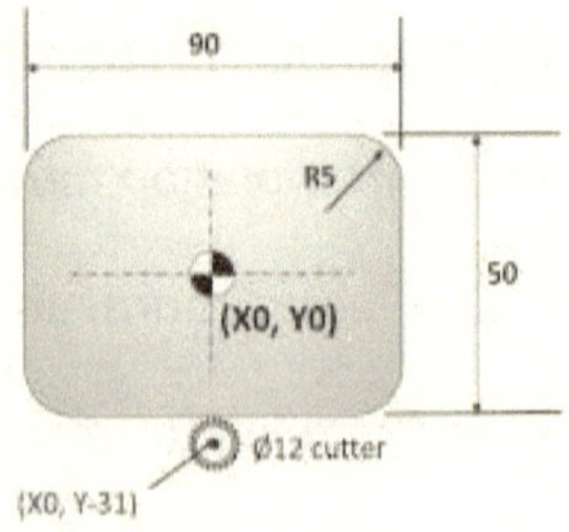

O1001 (Rectangular Boss - Manual Offset)

N10 G90 G54 G00 X0 Y0 (Work coordinate system)

N20 T03 M06 (Load 12mm end mill)

N30 S2000 M03 (Spindle on, 2000 RPM)

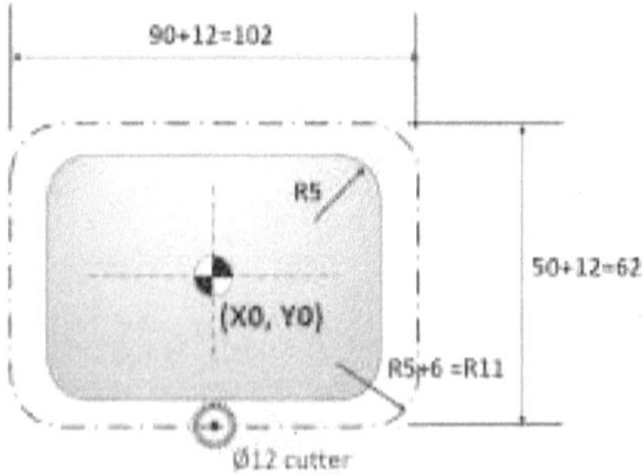

N40 G00 X0 Y-35.0 (Approach position - outside pocket)

N50 G00 Z5.0 (Rapid to clearance height)

N60 G01 Z-10.0 F100 (Plunge to pocket depth)

(Manual compensation with lead-in move)

N70 G01 Y-31.0 F200 (6mm added to Y coordinate 25mm)

(All coordinates manually calculated)

N80 G01 X40.0 Y-31.0 (Bottom edge)

N90 G03 X51.0 Y-20.0 R11.0 (Compensated Corner radius)

N100 G01 X51.0 Y20.0 (Right edge)

N110 G03 X40.0 Y31.0 R11.0 (Corner radius)

N120 G01 X-40.0 Y31.0 (Top edge)

N130 G03 X-51.0 Y20.0 R11.0 (Corner radius)

N140 G01 X-51.0 Y-20.0 (Left edge)

N150 G03 X-40.0 Y-31.0 R11.0 (Corner radius)

N160 G01 X0 Y-31.0 (Return to start)

N170 G01 Y-35.0 (retract)

2.16.9 Program with CRC (Direct Part Geometry – Simple):

O1002 (Rectangular Boss - With CRC)

N10 G90 G54 G00 X0 Y0 (Work coordinate system)

N20 T03 M06 (Load 12mm end mill - radius in offset table)

N30 S2000 M03 (Spindle on, 2000 RPM)

N40 G00 X0 Y-35.0 (Approach position - outside pocket)

N50 G00 Z5.0 (Rapid to clearance height)

N60 G01 Z-10.0 F100 (Plunge to pocket depth)

(Activate compensation with lead-in move)

N70 G01 **G42** Y-25.0 F200 (Comp right, move to start)

(Program exact part geometry - no offset math!)

N80 G01 X40.0 Y-25.0 (Bottom edge)

N90 G03 X45.0 Y-20.0 R5.0 (Corner radius)

N100 G01 X45.0 Y20.0 (Right edge)

N110 G03 X40.0 Y25.0 R5.0 (Corner radius)

N120 G01 X-40.0 Y25.0 (Top edge)

N130 G03 X-45.0 Y20.0 R5.0 (Corner radius)

N140 G01 X-45.0 Y-20.0 (Left edge)

N150 G03 X-40.0 Y-25.0 R5.0 (Corner radius)

N160 G01 X0 Y-25.0 (Return to start)

(Cancel compensation with lead-out move)

N170 G01 **G40** Y-35.0 (Comp cancel, retract)

N180 G00 Z50.0 (Rapid to safe height)

N190 M05 (Spindle off)

N200 M30 (Program end)

Program can be made further simplified by removing repeated coordinates

2.16.10 Program with CRC (no repeated coordinates):

O1002 (Rectangular Boss - With CRC)

N10 G90 G54 G00 X0 Y0 (Work coordinate system)

N20 T03 M06 (Load 12mm end mill - radius in offset table)

```
N30 S2000 M03 (Spindle on, 2000 RPM)
N40 G00 X0 Y-35.0 (Approach position - outside pocket)
N50 G00 Z5.0 (Rapid to clearance height)
N60 G01 Z-10.0 F100 (Plunge to pocket depth)
(Activate compensation with lead-in move)
N70 G01 G42 Y-25.0 F200 (Comp right, move to start)
(Program exact part geometry - no offset math!)
N80 G01 X40.0 (Bottom edge)
N90 G03 X45.0 Y-20.0 R5.0 (Corner radius)
N100 G01 Y20.0 (Right edge)
N110 G03 X40.0 Y25.0 R5.0 (Corner radius)
N120 G01 X-40.0 (Top edge)
N130 G03 X-45.0 Y20.0 R5.0 (Corner radius)
N140 G01 Y-20.0 (Left edge)
N150 G03 X-40.0 Y-25.0 R5.0 (Corner radius)
N160 G01 X0 (Return to start)
(Cancel compensation with lead-out move)
N170 G01 G40 Y-35.0 (Comp cancel, retract)
N180 G00 Z50.0 (Rapid to safe height)
N190 M05 (Spindle off)
```

N200 M30 (Program end)

Key Advantages Demonstrated:

- Coordinates match part drawing exactly (45, 25, 5 mm values)
- No manual offset calculations required
- Changing to a Ø10 mm or Ø16 mm tool requires only updating the tool offset table
- Program remains valid even as tool wears (adjust offset value only)

2.16.11 Example 2: Circular Boss (External Profile)

Part Requirements:

- Circular boss: Ø50 mm
- Height above base: 5 mm
- Material: Steel

Tooling:

- Tool: Ø16 mm end mill (radius = 8 mm)
- Tool number: T05

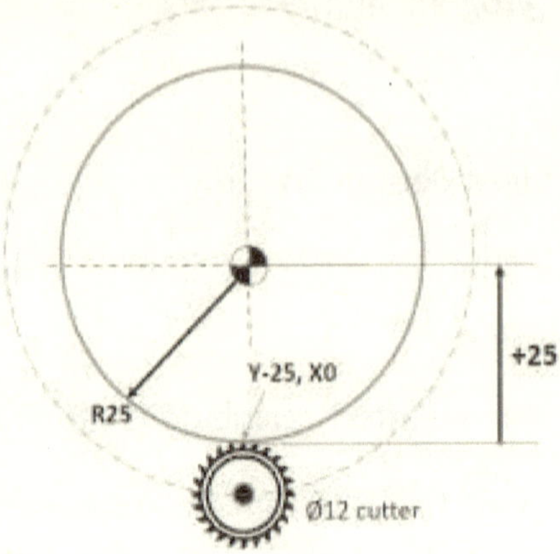

Program:

O1003 (Circular Boss - External Profile)

N10 G90 G54 G00 X0 Y0

N20 T05 M06 (Load 16mm end mill)

N30 S1500 M03

N40 G00 X0 Y-35.0 (Start position outside part)

N50 G00 Z2.0

N60 G01 Z-5.0 F80 (Cut to final depth)

(Lead-in move with compensation)

N70 G01 G41 Y-25.0 F150 (Comp left for CCW travel)

(Program exact 50mm diameter circle)

N80 G02 J25.0 (Full 360° arc)

(Lead-out and cancel)

N90 G01 G40 Y-35.0 (Cancel compensation)

N100 G00 Z50.0

N110 M05

N120 M30

Why G41 for this external profile?

- We're traveling clockwise (G02) around the outside
- Looking down at the part, the tool needs to be on the left side of the programmed path
- This pushes the tool outward by the radius, creating the correct 50 mm diameter

2.16.12 Practical Notes and Critical Pitfalls

2.16.12.1 Lead-in/Lead-out Moves Required

Rule: Always activate and cancel CRC with a straight-line approach move—never directly at the part edge.

Why? The control needs at least one full block of motion to calculate the offset vector perpendicular to the path. Starting compensation at a corner or arc can cause:

- Alarm conditions (typical error: "CRC vector calculation error")
- Gouging or tool collision
- Unpredictable offset direction

Best Practice:

(CORRECT - Straight lead-in)

N50 G00 X0 Y-40.0 (Outside the part)

N60 G01 G41 Y-30.0 F200 (Activate on straight move)

N70 G01 X20.0 Y-30.0 (Continue to profile)

(INCORRECT - Activating at a corner)

N50 G00 X0 Y-30.0

N60 G01 G41 X20.0 Y-20.0 (Diagonal - may alarm!)

Lead-in distance recommendation: 2 to 3mm more than tool radius for safe calculation.

5.2 Tool Table Accuracy is Critical

The tool's actual radius must be correctly entered in the offset table. A mismatch directly causes dimensional errors:

Example Error:

- Actual tool radius: 8.0 mm
- Entered in offset table: 6.0 mm
- Result: Pocket will be oversized by 4 mm on diameter (2 mm per side)

Verification procedure:

1. Measure tool diameter with a micrometer or tool presetter

2. Calculate radius (diameter ÷ 2)
3. Enter in tool offset table with appropriate precision (typically 0.001 mm)
4. Re-verify after tool changes or sharpening

Tool wear consideration: As tools wear, the effective radius decreases. Update the offset value to maintain dimensional accuracy without reprogramming.

2.16.13 Climb vs. Conventional Milling (Down milling vs. Up milling)

The choice of G41 or G42 also depends on whether you want climb milling or conventional milling around the profile.

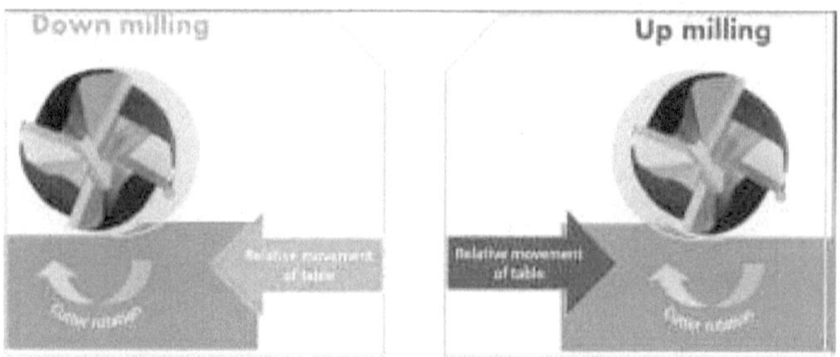

Climb Milling / Down milling (preferred for most CNC work):

- Chip thickness decreases from entry to exit
- Better surface finish
- Less tool deflection

- Increased tool life because most of the heat is carried away by the chips
- Requires rigid machine and sharp tools

Conventional Milling / Up milling:

- Chip thickness increases from entry to exit
- Better for worn tools or less rigid setups
- Creates more heat

Example: For external profile, traveling counterclockwise:

- G41 = climb milling (tool on left, cutting on right side)
- G42 = conventional milling (tool on right, cutting on left side)

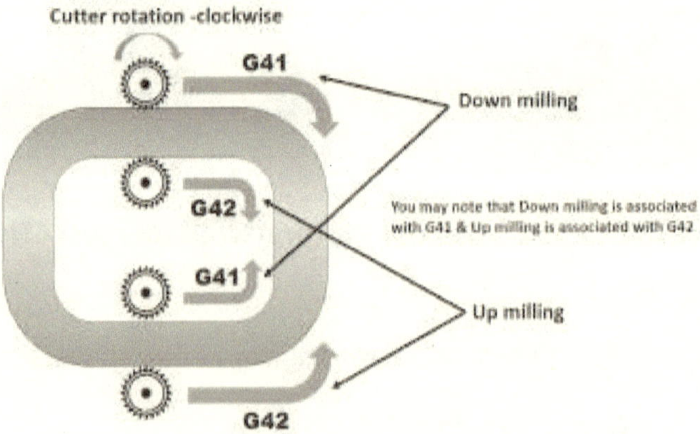

2.16.14 Common Beginner Errors

Error 1: Applying CRC on first move that is diagonal or arc-shaped

(WRONG)

N50 G01 G41 X20.0 Y30.0 (Diagonal activation - alarm likely)

(CORRECT)

N50 G01 G41 X20.0 Y0 (Straight line first)

N60 G01 X20.0 Y30.0 (Then diagonal)

Error 2: Forgetting to cancel compensation

N100 G01 X-40.0 Y-25.0

N110 G01 X0 Y-25.0 (End of profile)

N120 G00 Z50.0 (Forgot G40 - still compensating!)

N130 G00 X0 Y0 (Unexpected offset - possible collision)

Error 3: Tool radius larger than corner radius

If the tool radius exceeds the programmed corner radius, the control cannot generate a valid offset path. This causes an alarm: "CRC radius too large for corner."

Solution: Use a smaller tool or increase corner radii in the part design.

Error 4: Rapid moves while compensation is active

Some controls allow G00 with active compensation, but this can be dangerous:

(RISKY)

N80 G01 G41 X20.0 Y20.0

N90 G00 X50.0 Y50.0 (Rapid with comp active - possible gouge)

(SAFER)

N80 G01 G41 X20.0 Y20.0

N90 G01 X50.0 Y50.0 F500 (Linear move at high feed)

Best practice: Use only G01/G02/G03 while compensation is active.

2.16.15 Inside vs. Outside Corners

The control handles inside and outside corners differently:

Outside corners: Tool decelerates, possibly pauses to maintain path accuracy **Inside corners:** Control may need to insert a small arc to avoid gouging (automatic on most modern controls)

Sharp inside corners (<90°): May cause interference. The control will alarm if the offset path causes the tool to "double back" on itself.

2.16.16 Why Cutter Compensation is Indispensable

Accuracy

Ensures the tool cuts the intended profile with precision. The control performs complex geometric calculations far more accurately than manual methods.

Flexibility

Tool changes don't require reprogramming. Replace a worn Ø12 mm tool with a fresh Ø12 mm tool (or even a Ø10 mm tool in a pinch), update the offset table, and continue production without editing a single line of code.

Productivity

Speeds up setup and reduces programming errors. A part that might take hours to program manually can be coded in minutes using CRC.

Standardization

Programs remain consistent and readable regardless of tool wear, substitutions, or different machine configurations. This is especially valuable in production environments with multiple machines.

Tool Wear Compensation

As cutting tools wear, their effective diameter decreases. With CRC, simply measuring the worn tool and updating the offset value maintains part accuracy without any program changes.

Design Flexibility

Engineering changes to part dimensions can often be accommodated by modifying the CNC program geometry directly, without recalculating offset paths.

Real-World Impact

Consider a production scenario:

Without CRC:

- Program a batch of parts with Ø12 mm tool
- Tool breaks mid-batch
- Only Ø10 mm tool available as replacement
- Result: Must reprogram entire job (30+ minutes downtime) OR scrap remaining parts

With CRC:

- Tool breaks mid-batch
- Replace with Ø10 mm tool
- Update offset table: Change radius from 6.0 to 5.0 mm
- Result: Resume production immediately (2-minute downtime)

This single scenario can save thousands of dollars in a production environment.

2.16.17 Advanced Considerations

Multiple Compensation Zones

Some complex parts require switching between G41, G42, and G40 multiple times within a single program. Each transition must use proper lead-in/lead-out moves.

Compensation in 3D Machining

While CRC primarily applies to 2D contouring, some 5-axis machines extend the concept to 3D surface machining with tool length compensation working in conjunction with radius compensation.

2.16.18 Simulation and Verification

Always verify CRC programs using CNC simulation software before running on the machine. Visual verification shows:

- Correct offset direction
- Proper lead-in/lead-out paths
- Absence of gouges or tool interference
- Actual cut material vs. programmed geometry

2.16.19 Control-Specific Behaviors

Different CNC controls (Fanuc, Siemens, Haas, Mazak) may have slight variations in how they handle:

- Minimum lead-in distance requirements
- Sharp corner algorithms
- Error detection and reporting
- Lookahead calculations

Always consult your specific machine's programming manual.

2.16.20 Key Takeaways

✅ **Cutter Radius Compensation (CRC) is not just a convenience—it's a necessity for accurate, efficient CNC machining.**

✅ **Master the three codes:**

- **G41** = Compensation Left
- **G42** = Compensation Right
- **G40** = Cancel Compensation

✅ **Program part geometry directly** from engineering drawings—let the control handle offset mathematics.

✅ **Always use lead-in and lead-out moves** to activate and cancel compensation safely.

✅ **Maintain accurate tool offset tables**—dimensional accuracy depends entirely on correct radius values.

✅ **Visualize tool position** relative to path direction when choosing G41 vs. G42.

✅ **Verify with simulation** before cutting to catch errors in a safe virtual environment.

By mastering CRC, machinists gain the ability to program directly from part geometry while maintaining flexibility and complete control over dimensional accuracy. This fundamental skill separates proficient CNC programmers from beginners and is essential for efficient, error-free production machining.

2.17 Optional Chamfering and Corner R in Fanuc

In CNC machining, sharp corners between linear toolpath segments are often undesirable due to stress concentration, burr formation, or the need for smooth transitions. Fanuc controls provide a simple and powerful method to automatically create **chamfers (C)** or **corner radii (R)** between two consecutive linear moves—without the need for additional G02 or G03 commands.

2.17.1 Principle of Operation

When two straight lines meet at an angle, Fanuc can automatically insert a **chamfer or corner radius** at their intersection if the second block includes a **C** or **R** address.
The control calculates the intersection point of the two lines and modifies the toolpath accordingly.

2.17.2 Syntax

N110 G01 X__ Y__ R__ F_;

N120 G01 X__ Y__ ;

or

N110 G01 X__ Y__ C__ ;

- **R** → specifies the corner radius value.
- **C** → specifies the chamfer length along each leg.

A block that specifies a chamfer (**C**) or corner radius (**R**) must be immediately followed by another block containing a move command using **linear interpolation (G01)** or **circular interpolation (G02 or G03)**.

2.17.3 Example – Corner Radius

O1002 (Rectangular Boss - With CRC)

N10 G90 G54 G00 X0 Y0 (Work coordinate system)

N20 T03 M06 (Load 12mm end mill - radius in offset table)

N30 S2000 M03 (Spindle on, 2000 RPM)

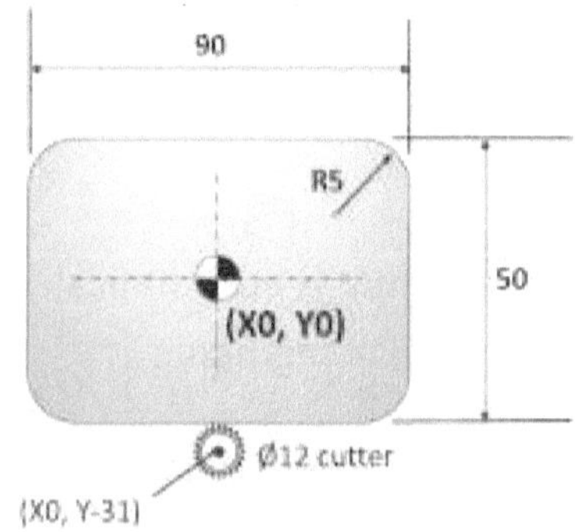

N40 G00 X0 Y-35.0 (Approach position - outside pocket)

N50 G00 Z5.0 (Rapid to clearance height)

N60 G01 Z-10.0 F100 (Plunge to pocket depth)

(Activate compensation with lead-in move)

N70 G42 Y-25.0 F200 (Comp right, move to start)

(Program exact part geometry - no offset math!)

N80 G01 X45.0 R5.0 (Automatic corner rounding)

N90 Y25.0 R5.0

N100 X-45.0 R5.0

N110 Y-25.0 R5.0

N120 X0 (Return to start)

(Cancel compensation with lead-out move)

N130 G40 Y-35.0 (Comp cancel, retract)

N140 G00 Z50.0 (Rapid to safe height)

N150 M05 (Spindle off)

N160 M30 (Program end)

Result:
A corner radius of 5 mm is automatically blended between the two linear segments.

2.17.4 Example – Chamfering

O1002 (Rectangular Boss)

N10 G90 G54 G00 X0 Y0 (Work coordinate system)

N20 T03 M06 (Load 12mm end mill - radius in offset table)

N30 S2000 M03 (Spindle on, 2000 RPM)

N40 G00 X0 Y-35.0 (Approach position - outside pocket)

N50 G00 Z5.0 (Rapid to clearance height)

N60 G01 Z-10.0 F100 (Plunge to pocket depth)

(Activate compensation with lead-in move)

N70 G42 Y-25.0 F200 (Comp right, move to start)

(Program exact part geometry - no offset math!)

N80 G01 X45.0 C5.0 (Automatic corner chamfering)

N90 Y25.0 C5.0

N100 X-45.0 C5.0

N110 Y-25.0 C5.0

N120 X0 (Return to start)

(Cancel compensation with lead-out move)

N130 G40 Y-35.0 (Comp cancel, retract)

N140 G00 Z50.0 (Rapid to safe height)

N150 M05 (Spindle off)

N160 M30 (Program end)

Result:
A chamfer of 5 mm is automatically created at the corners.

2.17.5 Rules and Limitations

- A block that specifies a chamfer (**C**) or corner radius (**R**) must be immediately followed by another block containing a move command using **linear interpolation (G01)** or **circular interpolation (G02 or G03)**.
- The radius or chamfer must be small enough that it fits within the two intersecting lines.
- Only one of C or R may be specified per corner.

2.17.6 Practical Tips

- Chamfer and corner radius are Ideal for creating simple edge breaks on pockets, profiles, and rectangular paths.
- Always perform a **dry run** to ensure the chamfer or radius fits geometrically at internal corners.

2.18 Canned Cycles

2.18.1 Introduction — Automating Repetitive Operations

In CNC machining, many operations—drilling, boring, tapping, reaming, and rough turning—repeat the same motion sequence at multiple coordinates. Writing every rapid, feed, dwell, and retract step for each feature leads to long programs, error-prone edits, and poor readability.

Canned cycles encapsulate these sequences into a single G-code call. You invoke the cycle once, and then change only the X–Y positions for subsequent features. This reduces program length, simplifies edits, eliminates repetitive mistakes, and keeps the process logic consistent across all features.

2.18.2 What Is a Canned Cycle?

A canned cycle is a controller-resident subroutine. When you call a cycle such as **G81**, **G83**, or **G76**, the control executes a predefined sequence:

- Rapid positioning to the X–Y location
- Rapid approach to the R-plane (reference plane)
- Feed from the R-plane to the programmed Z depth
- Optional actions at depth: dwell, spindle stop, peck retract, spindle reversal
- Retraction to either the initial Z plane (**G98**) or the R-plane (**G99**)

Though details vary across cycles, the overall structure and syntax remain consistent.

2.18.3 Symbols in figures

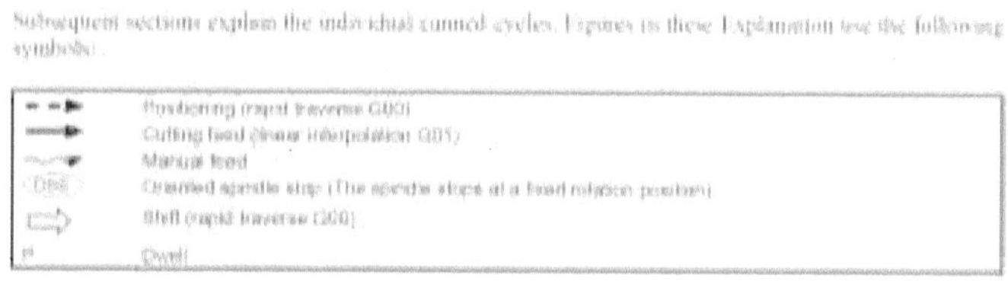
Subsequent sections explain the individual canned cycles. Figures in these explanation use the following symbols:

Positioning (rapid traverse G00)
Cutting feed (linear interpolation G01)
Manual feed
Oriented spindle stop (The spindle stops at a fixed rotation position)
Shift (rapid traverse G00)
Dwell

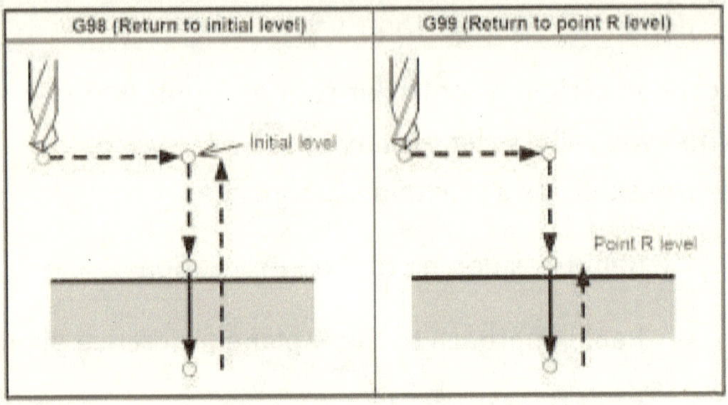

2.18.4 General Syntax and Address Words

Standard Format:

G8x X__ Y__ Z__ R__ Q__ P__ F__ K__

Address Meanings

Address	Description	Notes
X, Y	Feature location	Absolute (**G90**) or incremental (**G91**)
Z	Final cutting depth	Typically, negative relative to work offset

Address	**Description**	**Notes**
R	Reference plane	Clearance above the surface
Q	Peck depth	Used in **G83**, **G73**
P	Dwell time	Fanuc uses milliseconds
F	Feed rate	mm/min or in/min
K	Repeats or passes	Common in turning cycles

Modal Behavior:

X, Y, Z, R, and F remain active until changed. Usually, only X and Y vary between holes; adjust Z or F when depth or feed must differ.

2.18.5 Milling Cycles (Fanuc/ISO)

2.18.5.1 Drilling

Cycle	Function	Motion Characteristics	Typical Use
G73	High-speed peck	Short chip-break retracts, no full withdraw	Shallow holes, free-cutting materials
G81	Standard drilling	Feed to depth, rapid retract	General use, <3×D
G82	Drill with dwell	Feed, dwell, rapid retract	Spot faces, counterbores
G83	Deep-hole peck	Full retract to R-plane between pecks	Deep or blind holes, chip evacuation

Notes: G73 is faster but retains chips in the hole; G83 is safer for deep or blind holes and tough materials.

2.18.5.2 high speed drilling cycle G73

G98 G73 X20. Y15. Z-20. R2. Q3. D1. F100

G91 X50. K10

G90 G80

Explanation:

X20. Y15. = First Hole position

Z-20. = Final depth of the hole

R2. = Rapid approach height

Q3. = Peck depth increment

D1. = Retraction amount

F = Feed rate

Drill movements

Rapid position to first hole (X20. Y15.)

Rapid to **Z=2 mm** (R-plane).

Feed 3 mm to **Z=-3**.

Retract up **1 mm above** that point (Z=-2).

Feed next peck 3 mm to Z=-6, retract again to Z=-5, etc.

Continue until Z=-20.

Rapid back to **R-plane (Z=2)**, then to **initial Z** because of G98.

Then move by 50mm incremental distance in X direction and drill.

Repeat this for 9 more X locations (by moving 50mm each)

Note: Please check with your machine documentation if R is an absolute value or incremental

2.18.5.3 Tapping

Cycle	Type	Synchronization	Notes
G84	Rigid tapping (RH)	Spindle encoder synchronized to Z	Use **G98** for full retract
G74	Left-hand tapping	Spindle reversed	LH threads
G84 with floating holder	Non-rigid	Mechanical compensation	For older machines without rigid tapping

Feed rate Formula:

F=pitch × RPM

Example: M10 ×1.5 at 500 rpm → F750 mm/min

2.18.5.4 Boring

Cycle	Spindle Action	Retract Mode	Precision	Application
G85	Continuous rotation	Feed in and out	±0.02 mm	Semi-finish
G86	Stop at depth	Rapid retract while stopped	±0.05 mm	Rough boring
G76	Orient + offset	Controlled offset retract	±0.005 mm	Fine boring
G87	Reverse/back boring	Retraction	—	Back-face counterbores

Cycle	Spindle Action	Retract Mode	Precision	Application
		reverse side		
G88	Stop + manual retract	Operator controlled	—	Special fixtures
G89	Continuous + dwell	Feed retract after dwell	±0.01 mm	High precision

Fine boring (G76) sequence: Feed to depth → orient spindle → offset tool → retract clear → recenter at R-plane.

2.18.5.5 Fine boring Cycle G76

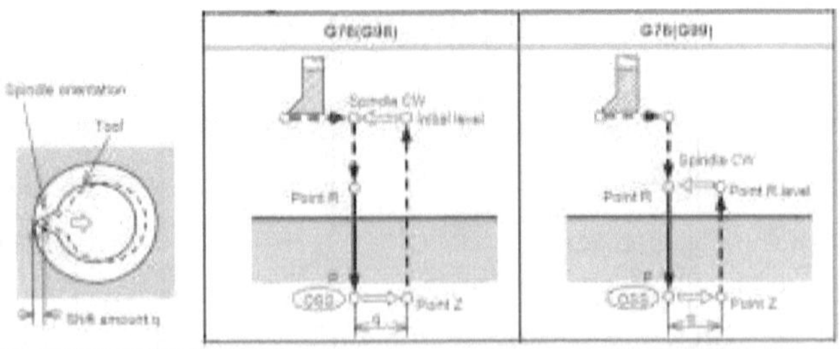

OSS – Oriented Spindle Stop

2.18.5.6 Example for Fine boring Cycle G76

G90 G17 G40 G80

T1 M6 (Fine-boring head)

G54 G0 X0. Y0. (Hole position)

S2200 M3

G43 H1 Z100.

G98 G76 X0. Y0. Z-20. R2. I0.2 P200 F80

G80

G0 Z100.

M5

Explanation:

- G76 – Fine boring cycle (shift at bottom after spindle orient)
- X0. Y0. – Hole position
- Z-20. – Final depth (mm)
- R2. – R-plane (rapid approach height)
- F80 – Cutting feed (mm/min)
- I0.2 – **Shift amount** at the bottom (radial tool-center shift, mm).

- P200 – **Dwell at the bottom** after shifting (milliseconds on many controls).
- G98 – After the cycle finishes, return to the **initial Z**; use G99 to return only to the R-plane.
- G80 – Cancel canned cycle.

Motion sequence (per hole)

1. Rapid to R (Z=+2.0).
2. Feed to Z=-20.0 at F80.
3. Spindle **stops and orients**
4. Control applies a **radial micro-shift** of I=0.2 mm (tip away from bore)
5. **Dwell** P=200 ms
6. Retract to R (and with G98, continue to the initial Z).

   When the cutting tool reaches the bottom of the hole, the spindle automatically stops and orients to a fixed angular position. The boring tool is then shifted slightly in the direction **opposite to the tool nose** and retracted.

7. This motion prevents the tool from rubbing against the freshly machined surface during withdrawal. As a result, the inner wall of the bore remains smooth and undamaged, ensuring **high precision and excellent surface finish**.

2.18.6 Practical Programming Examples

2.18.6.1 Milling: Bolt Circle Drilling (G83)

```
%
O1001
G90 G17 G21 G40 G80 G54
T05 M06
S1500 M03
G00 X0 Y0 Z100.
G98 G83 X0 Y30. Z-25. R2. Q5. F120
X23.511 Y19.021
X9.270 Y-4.635
X-9.270 Y-4.635
X-23.511 Y19.021
G80
G00 Z100. M09
M30
%
```

Q5 = 5 mm peck. R2 = 2 mm clearance. G98 returns to initial Z between holes.

2.18.6.2 Milling: Tapping Three Holes (G84)%

```
O1002
G90 G17 G54 G40 G80
T07 M06
S400 M03
G43 Z50. H07
G00 X10. Y10. M08
G98 G84 X10. Y10. Z-20. R2. F500
X40. Y10.
X25. Y35.
G80
M05 M09
G00 Z50.
M30
```

F = pitch × rpm = 1.25 × 400 = 500 mm/min.

2.18.7 Modal Behavior and Cancellation

A canned cycle remains active until canceled with **G80**, replaced by another cycle, or terminated with **M02/M30**.

Return Modes:

- **G98**: Returns to initial Z plane (safe when clearance varies).
- **G99**: Returns only to R-plane (faster, requires proven safety).

2.18.8 Techniques and Best Practices

R-Plane Safety

- Keep 1–2 mm above the highest local surface.
- Add 5–10 mm margin for castings/forgings.
- Use G98 where clamp or height variation exists.

Subprograms with Cycles
Group hole lists in a subprogram; apply coordinate rotation or mirroring for repeated patterns.

Peck Depth (G83)

- Q = 1–2 × drill diameter (typical).
- Reduce Q for stainless or titanium; increase for aluminum and plastics.

Dwell Time (G82/G89)

- Spot faces: 0.5–1.0 s.
- Counterbores: up to 2.0 s.
- Too long a dwell may harden the surface.

Checklist
R above surface; Z negative; realistic F; G80 cancel; G98 safe retract; F = pitch × rpm for tapping; simulate before running.

2.18.9 Troubleshooting

Symptom	Likely Cause	Remedy
Crash at first hole	R below surface	Raise R; verify work offset
Broken taps	Feed ≠ pitch × rpm	Correct feed; confirm rigid tap mode
Rough bores	Feed too high or dwell too short	Reduce feed; add dwell
Chips packing	Peck too large	Reduce Q; improve coolant flow
Tool hits clamp	Used G99	Use G98 for full retract

Symptom	Likely Cause	Remedy
Cycle repeats unexpectedly	Missing G80	Always end with G80
Torn threads	First pass too deep (G76)	Reduce Q (first-pass depth)

Always confirm parameters using the controller manual supplied with your machine.

2.18.10 Key Takeaways

Canned cycles let the machine control manage repetitive motion details while the programmer focuses on tooling and strategy. Mastery of these canned cycles shortens code, prevents errors, and improves program clarity and maintainability.

2.18.11 Practice and References

Exercises

- Drill a 4×6 rectangular pattern using G81.
- Tap an 8-hole bolt circle using G84.
- Program a deep-hole routine with G83 at 10:1 L/D.
- Fine-bore a bearing seat using G76 (±0.005 mm).

References

- *Fanuc Series 0i/30i/31i/32i Programming Manual (B-63944EN)*
- *ISO 6983 (G-code standard)*

2.19 CNC Turning Programming

2.19.1 Introduction: The Logic of CNC Programming

Welcome to the world of CNC programming. In modern manufacturing, the ability to translate a component drawing into a set of precise instructions for a machine is a cornerstone of efficiency, accuracy, and repeatability. The logic behind this process is surprisingly intuitive. Think of a simple "join the dots" game: you are given a series of numbered points, and by connecting them in the correct sequence with lines, a clear picture emerges.

CNC programming operates on a very similar principle. We define specific points on a workpiece using coordinates, and then we use programming commands to instruct a cutting tool on how to connect those points—either with a straight line or a perfect arc—to create the final part geometry.

The objective of this training module is to provide you with a foundational understanding of the core programming commands for CNC turning centers. By the end of this session, you will be able to read, understand, and write basic programs, transforming a digital blueprint into a physical component. Our first step is to understand the "map" on which we will draw our tool path: the machine's coordinate system.

2.19.2 Core Concepts: Defining the Machining Environment

Before any movement can be programmed, an operator must understand the machine's fundamental "map"—its axis system and coordinates. This framework is the absolute foundation for all CNC programming, as every command for movement is relative to this defined space. Getting this right is the first and most critical step toward successful machining.

2.20 Axis Naming Convention

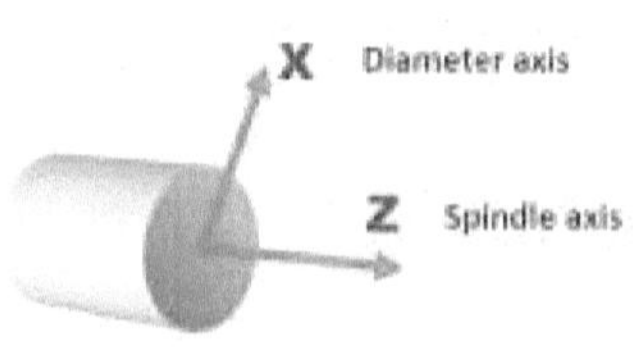

For a CNC turning center (or lathe), the axis naming convention is standardized and simple to remember:

- The axis parallel to the main spindle is **always the Z-axis**.

- The other primary axis, which controls the diameter of the workpiece, is the **X-axis**.

2.21 The Coordinate System

2.21.1 Four Quadrants

CNC turning operations are plotted on a four-quadrant coordinate system defined by the intersection of the X and Z axes. Understanding these quadrants is key to visualizing the tool's position relative to the workpiece.

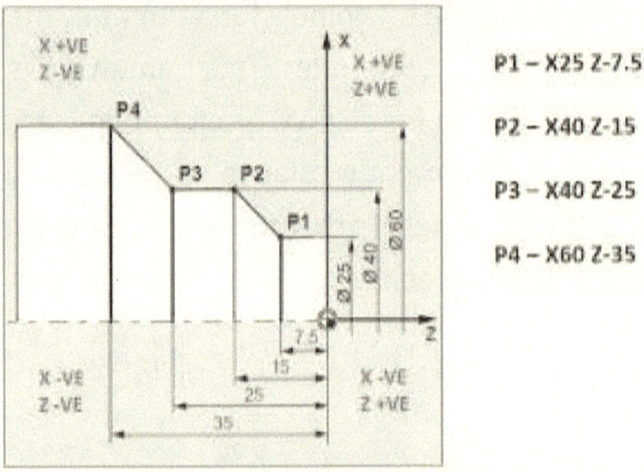

Quadrant	Coordinate Properties
First	X is Positive (+) / Z is Positive (+)
Second	X is Positive (+) / Z is Negative (-)
Third	X is Negative (-) / Z is Negative (-)
Fourth	X is Negative (-) / Z is Positive (+)

In the vast majority of turning applications, programming occurs in the **second quadrant (X+, Z-)**. This is because the program's zero point (Z0) is typically set at the finished face of the workpiece. As the tool moves into the material toward the chuck, its Z-axis position becomes negative. Since the X-axis represents the diameter, it is almost always a positive value.

2.21.2 Defining Tool Path Points

Coordinates directly translate to the physical dimensions of the part. In turning, the X value represents a specific **diameter**, and the Z value represents a specific **length** or position along the spindle axis.

Consider the following points for a sample workpiece:

- **P1: X25, Z-7.5**
    - This point defines a location at a diameter of 25mm and a distance of 7.5mm into the part from the Z0 face.
- **P2: X40, Z-15**
    - This point is at a diameter of 40mm and a length of 15mm from the face.
- **P3: X40, Z-25**
    - This point is at a diameter of 40mm and a length of 25mm from the face.
- **P4: X60, Z-35**

- This point is at a diameter of 60mm and a length of 35mm from the face.

Now that we understand how to define locations, we can learn the language used to command the machine to move between them.

2.22 The Language of Movement: G-Codes and M-Codes

The primary command language for any CNC machine is comprised of G-codes and M-codes. These simple alphanumeric commands form the building blocks of every program. It is crucial to understand their distinct roles: G-codes tell the machine *how and where to move*, while M-codes control the machine's auxiliary functions.

It is important to note that while these codes are largely standardized, the specific examples and conventions in this module are based on the **Fanuc control system**, one of the most common in the industry. Comprehensive G-code lists are added as annexure

2.22.1 G-Codes (Preparatory Functions)

G-codes are **preparatory functions**. They prepare the machine's control system for a specific type of action, primarily related to geometry and motion. This includes telling the machine what units to use, what measurement system to follow (absolute or incremental), and, most importantly, what kind of movement to perform. For Fanuc, there are 3 types of systems for G-codes namely A, B & C. By default, **G-code system A** will be active. In this book we will be describing the use of system A.

2.22.2 M-Codes (Miscellaneous Functions)

M-codes are **miscellaneous functions**, often called machine functions. These codes control the physical operations of the machine that are not related to axis movement. Common examples include:

- Spindle On / Off
- Spindle Direction (Clockwise / Counter-Clockwise)
- Coolant On / Off

2.22.3 Analysis of Key Motion G-Codes

The following G-codes are the fundamental commands for tool movement. Understanding their distinct purposes is essential for efficient and safe programming.

Code	Function	Operational Analogy & Purpose
G00	**Rapid Traverse**	**Analogy: Superbike.** This command is for maximum-speed, non-cutting movements. Its purpose is to position the tool quickly in the air to save valuable

		cycle time. Cutting should *never* occur during a G00 move.
G01	**Linear Interpolation**	**Analogy: Bicycle.** This command is for controlled, straight-line cutting movements at a specified feed rate. It is used for facing, turning straight diameters, and machining tapers.
G02	**Circular Interpolation Clockwise**	This command instructs the tool to move in a clockwise arc to create convex radii and other curved features.
G03	**Circular Interpolation Counter-Clockwise**	This command instructs the tool to move in a counter-clockwise arc to create concave radii and other curved features.

With these commands defined, we can now explore how they are applied to create common part geometries.

2.23 Programming Linear Movements (G01)

The G01 command, or Linear Interpolation, is the true workhorse of CNC turning. It is used to generate the most fundamental part geometries, including flat faces, straight outside diameters (OD), and tapers.

2.23.1 Facing Operations

A facing operation creates a flat, smooth surface on the end of the workpiece, typically at the Z0 position. Let's deconstruct a typical facing sequence for a part with a Ø25mm diameter.

Step 1: Positioning (G00 X27 Z0)

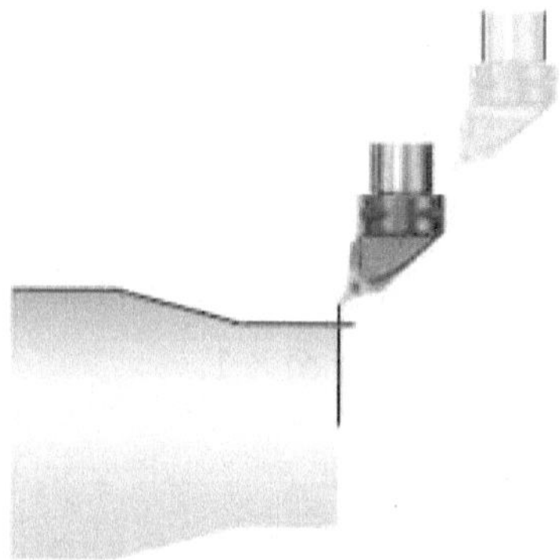

- This is a rapid positioning move. The tool is commanded to move quickly to X27, which is a safe position 2mm diametrically above the Ø25mm part. It also moves to Z0, aligning the tool tip with the face of the part, ready to begin the cut.

Step 2: Cutting (G01 X-1 F0.25)

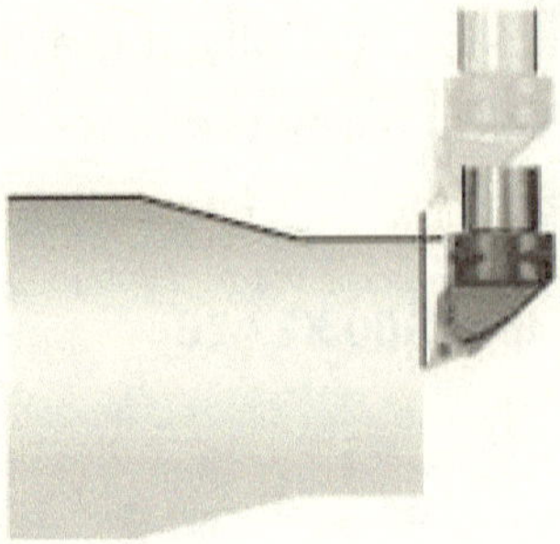

- This is the linear cutting move. The tool is instructed to move in a straight line along the X-axis to a position of X-1.
- **Analysis:** Facing is an X-axis-only command. Critically, we program to X-1 instead of X0. This small over-travel compensates for the tool's corner radius, ensuring that a small "pip" or nub is not left at the exact center of the workpiece.
- The F0.25 command defines the **feed rate**—the speed at which the tool advances during the cut.

2.23.2 OD Turning (Straight & Taper)

Turning an outside diameter profile involves a combination of straight and tapered movements, all controlled by the G01 command.

Step 1 (Straight Turn): G01 Z-7.5

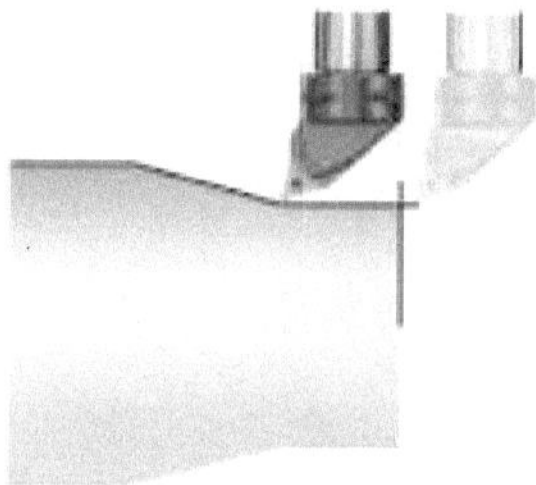

- After positioning the tool at the starting diameter (e.g., X25), a command with only a Z-axis endpoint will result in a straight turn. The tool moves parallel to the Z-axis, maintaining a constant diameter.

Step 2 (Taper Turn): G01 X40 Z-15

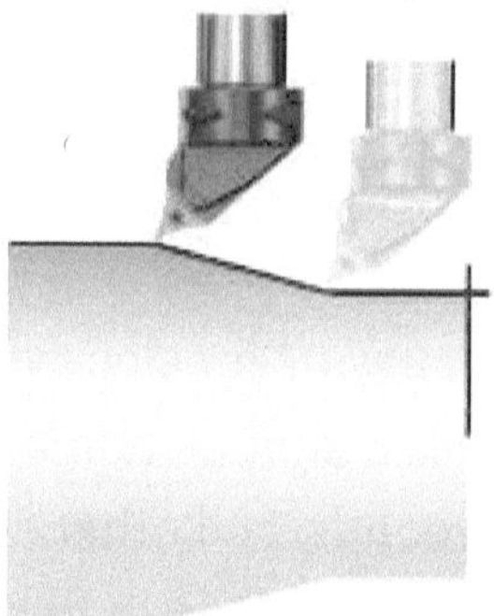

- To create a taper, we must define both the X and Z coordinates for the end point of the move simultaneously. The control system calculates the straight-line path between the start point and the end point (X40, Z-15), resulting in a precise angular cut.

Step 3 (Straight Turn): G01 Z-25

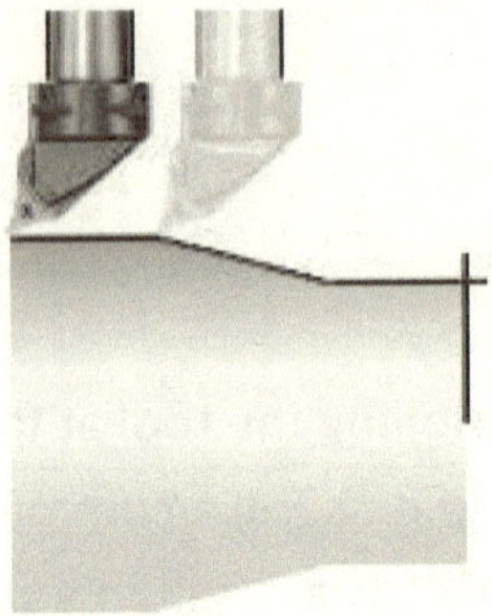

- Once the taper is complete, we can again issue a Z-axis-only command to resume a straight turn at the new diameter.

Creating shapes with straight lines is fundamental, but to machine more complex profiles, we must learn to create curves.

2.24 Programming Arcs (G02 & G03)

Circular interpolation (G02 for clockwise, G03 for counter-clockwise) is essential for machining profiles that include radii, rounds, and other curved features. There are two primary methods for programming an arc, and a proficient programmer should be familiar with both.

2.24.1 Method 1: Using the Radius (R) Parameter

The simplest method for defining an arc is to specify its end point coordinates (X and Z) and its radius value using the R parameter.

1. **Counter-Clockwise Arc:** G03 X60 Z-90 R50

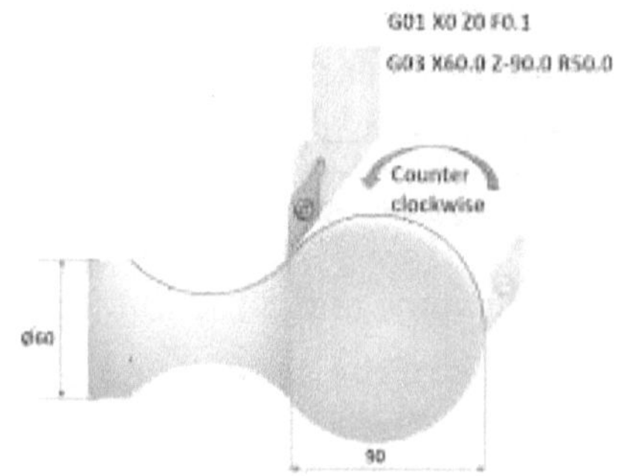

G03: Specifies a counter-clockwise movement.

X60 Z-90: Defines the end point of the arc.

R50: Defines the radius of the arc as 50mm.

2. **Clockwise Arc:** G02 X60 Z-150 R50

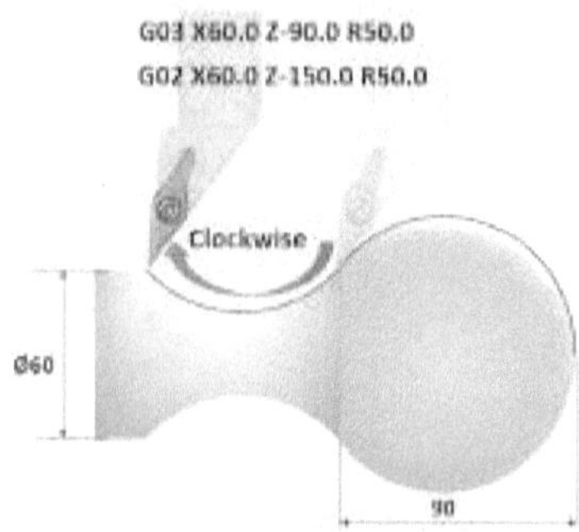

G02: Specifies a clockwise movement.

X60 Z-150: Defines the end point of the arc.

R50: Defines the radius of the arc as 50mm.

2.24.2 Method 2: Using I and K Parameters

An alternative and often more powerful method is to define the arc's center point using I and K parameters. The definition for these parameters is precise and must be remembered exactly.

"I and K are distance and direction from start point of the arc to center of the arc in X and Z directions."

It is critical to calculate these values *from the start point of the arc to its center*. I represents the incremental distance and direction in the X-axis, while K represents the incremental distance and direction in the Z-axis.

Counter-Clockwise Example: G03 X60 Z-90 I0 K-15

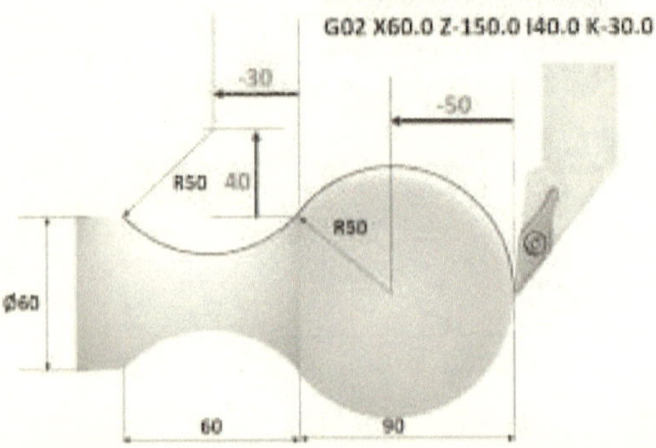

Analysis:

I0: This means there is no distance in the X-direction between the arc's start point and its center point.

K-15: This means the arc's center is located 15mm in the negative Z-direction from the arc's start point.

Clockwise Example: G02 X60 Z-150 I40 K-30

Analysis:

I40: The arc's center is located 40mm in the positive X-direction from the start point.

K-30: The arc's center is located 30mm in the negative Z-direction from the start point.

Both the R and I/K methods will successfully create the required arc. Proficiency in choosing the right method for a given situation comes with practice.

Conclusion and Next Steps

In this module, you have learned the essential building blocks of CNC turning programming. We began by defining the machining environment through axes and coordinates. We then explored the command language of G-Codes and M-Codes and, most importantly, learned to apply the primary motion commands for both linear (G01) and circular (G02/G03) tool paths.

The "join the dots" analogy remains a powerful mental model: you define the points with coordinates and connect them with the appropriate G-code commands to form the final part geometry.

2.25 Absolute and Incremental in Turning (Using U and W)

In CNC turning, tool movements can be specified in either **absolute** or **incremental** mode. The distinction lies in how the control interprets positional data relative to a reference point. Understanding this concept is essential for precise toolpath definition and error-free programming.

2.25.1 Absolute Programming (Using X and Z)

In **absolute programming**, each position is defined **from a fixed reference point**, typically the **work coordinate system origin** (set using G54, G55, etc.).
All coordinates are measured relative to this zero point.

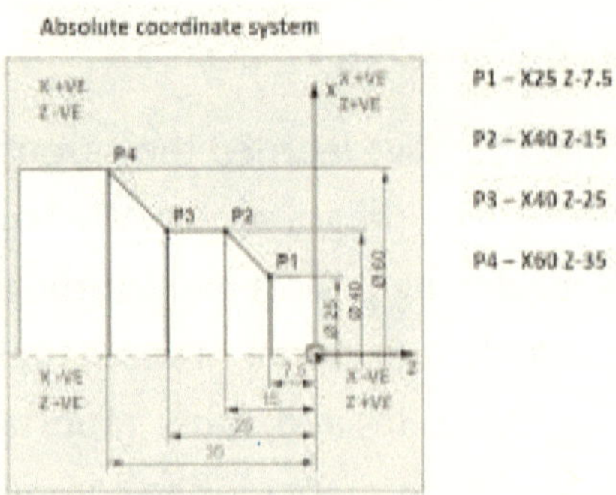

Syntax Example:

G00 X20.0 Z2.0

- The tool will move to an **absolute position** of 20.0 mm in X and 2.0 mm in Z.
- Every subsequent command uses the same origin as reference.

Key Features:

- Reference: Part zero (work offset origin)
- Axis commands: X and Z
- Easier to visualize part dimensions directly from drawing
- Preferred for most finish and production programs

2.25.2 Incremental Programming (Using U and W)

In **incremental programming**, each position is defined **relative to the tool's current position** rather than a fixed origin.
The axes use U (incremental X) and W (incremental Z) to indicate movement **distance** instead of position. Please note that this method is using **G-code system A**. G90 & G91 in this system are not for absolute / incremental. They are used for turning cycles.

Incremental coordinate system

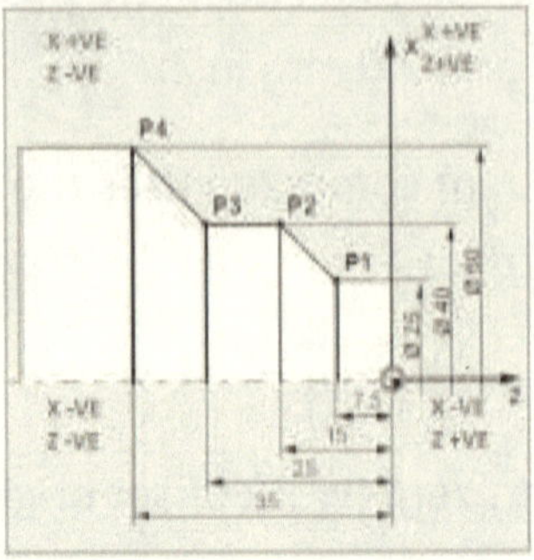

P0 – U25 Z0

P1 – U0 W-7.5

P2 – U15 W-7.5

P3 – U0 W-10

P4 – U20 W-10

Assumption:

Before starting P0, tool is in X0, Z0

Syntax Example:

G00 U2.0 W2.0

- The tool moves **1 mm away from part centerline (X direction)** and **2.0 mm away from the chuck (Z direction)** from its **current position**.
- After this move, the new position becomes the reference for the next incremental command.

Key Features:

- Reference: Previous tool position
- Axis commands: U and W
- Ideal for repetitive moves such as roughing passes, peck drilling, or pattern repeats
- Reduces the need for recalculating coordinates when repeating similar cuts

Example Comparison

Description	**Absolute Mode**	**Incremental Mode**
Move to a position 1.0 mm deeper in Z	G01 Z-21.0 F0.2	G01 W-1.0 F0.2
Move radially 0.2 mm closer to center	G01 X19.6 F0.1	G01 U-0.4 F0.1 *(since X is diameter-based)*

2.25.3 Notes for Turning Applications

- The **X-axis** in turning programs typically represents **diameter values**, so incremental moves using U are **twice the actual radial distance**.
- For instance, a command U-0.2 moves the tool **0.1 mm radially** toward the centerline.
- Switching between absolute and incremental modes can occur within the same program. It is even possible to mix both absolute and incremental in same program line as the addresses used are different (Example G01 Z-50. U0.1 F0.1)

2.25.4 Practical Tip

When roughing in cycles or macros, using **incremental (U, W)** makes it easy to define repeated depth cuts. For adjusting the taper observed in plane turning also, U can be used.
For finishing or coordinate-based toolpath control, **absolute (X, Z)** ensures dimensional accuracy aligned with the drawing.

2.26 Constant Cutting Speed Control (CSS) in CNC Turning

2.26.1 Introduction

In conventional lathes, the spindle rotates at a **fixed speed (revolutions per minute)** throughout the machining operation. However, this approach does not maintain a uniform cutting speed at the tool–workpiece interface. Since the workpiece diameter continuously decreases during turning, the **peripheral (surface) speed** also decreases if spindle RPM remains constant. This leads to inconsistent chip formation, variable surface finish, and fluctuating tool life.

To overcome this, **Constant Cutting Speed (CSS)** control is employed in CNC turning machines. CSS automatically adjusts the spindle RPM so that the **cutting speed (m/min or SFM)** remains constant, regardless of the changing workpiece diameter.

2.26.2 Principle of CSS Operation

The cutting speed Vc at the workpiece surface is given by:

$$V_c = \frac{\pi D N}{1000} \qquad N = \frac{1000 \times V_c}{\pi D}$$

Where:

- Vc = Cutting speed (m/min)
- D = Workpiece diameter (mm)
- N = Spindle speed (rev/min)

In CSS mode, the CNC control calculates N dynamically from the programmed cutting speed and the current workpiece diameter:

As the tool moves toward the center (smaller D), the control automatically increases the spindle RPM to maintain the commanded cutting speed Vc.

2.26.3 G-Code Commands

In most CNC controls (Fanuc, Siemens, Mitsubishi, etc.), CSS is activated using the **G96** command, and the system is returned to fixed-RPM mode using **G97**.

Command	Function	Example	Description
G96	Constant surface speed mode	G96 S180 M03	Cutting speed = 180 m/min; spindle rotates clockwise
G97	Cancel CSS (fixed RPM)	G97 S1200 M03	Spindle runs at 1200 RPM constant

Note: When G96 is active, the **S value** represents *cutting speed in m/min (metric)* or *surface feet per minute (imperial)* — **not RPM**.

2.26.4 Maximum Spindle Speed Limitation

When the tool approaches the workpiece center, the calculated spindle RPM theoretically tends to infinity. To prevent this, CNC systems allow programming a **maximum spindle speed limit** using the **G50** command.

Example:

G50 S2500 (Limit maximum spindle speed to 2500 RPM)

G96 S180 M03 (Activate CSS with 180 m/min cutting speed)

Here, even when the control tries to increase the speed as the diameter reduces, the spindle will not exceed **2500 RPM**.

2.26.5 Application Example

Objective: Do facing from Ø100 mm to Ø50 mm using 180 m/min cutting speed.

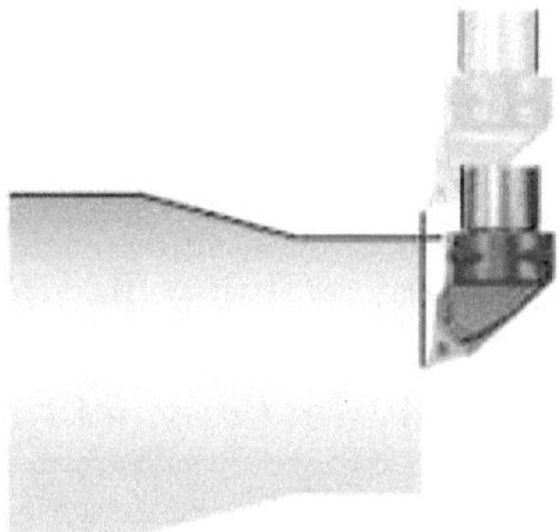

Program Segment:

G50 S4000

G96 S180 M03

G00 X102.0 Z2.0

G01 X-1. F0.3

G00 X102.0 Z2.0

G97 S1000 M05

- The machine calculates spindle RPM continuously based on actual diameter.

- At Ø100 mm: RPM ≈ 573 rev/min
- At Ø50 mm: RPM ≈ 1146 rev/min
- At Ø5 mm: RPM ≈ 11464 rev/min (here RPM gets locked at 4000)

This ensures constant cutting conditions and consistent tool load throughout the pass.

2.26.6 Advantages of CSS

- **Uniform Surface Finish:** Cutting speed remains consistent across varying diameters.
- **Improved Tool Life:** Constant heat generation and predictable wear.
- **Efficient Machining:** Maximizes productivity without manual speed adjustments.
- **Reduced Vibration:** Optimized cutting conditions minimize chatter and tool deflection.

2.26.7 Practical Considerations

1. **Always set G50** before activating CSS to avoid overspeed.
2. CSS is generally effective for **external turning, facing, and profiling** operations.
3. Avoid CSS in **threading, tapping, and drilling**, where fixed RPM synchronization is required.

4. Ensure proper spindle acceleration/deceleration characteristics are configured in the machine parameters.
5. CSS response can vary between machine models — check manufacturer recommendations.

2.26.8 Key take away

Constant Cutting Speed control (G96) in CNC turning maintains a **steady surface speed** regardless of changing diameters, resulting in better surface quality, stable tool wear, and improved productivity. By pairing **G96** with a **G50** limit and switching back to **G97** where required, machinists can achieve both **safety** and **performance optimization** in automated turning operations.

2.27 Tool Nose Radius Compensation (TNRC)

2.27.1 Introduction: The Secret to Perfect Parts

Have you ever wondered how a CNC machine produces perfectly sharp angles and smooth curves, even though the cutting tool itself has a rounded tip? The secret lies in a clever feature called Tool Nose Radius Compensation (TNRC).

Picture this: You program a sharp 45-degree chamfer, but when you measure the finished part, there's extra material at the corners. Or worse, you're cutting a beautiful radius, but the surface has irregular gouges. The culprit? An uncompensated tool nose radius.

In this section, we'll uncover:

- Why tool nose radius creates challenges in CNC turning
- How TNRC (G41/G42) solves those challenges automatically
- The foolproof rules for choosing between G41 (Left) and G42 (Right)
- Practical setup requirements with real-world examples
- Common mistakes and how to avoid them

By the end, you'll not only understand the concept but also be confident enough to apply TNRC in your own programs—and troubleshoot when things don't look right.

2.27.2 The Problem: Why We Can't Ignore the Tool's Shape

2.27.2.1 **Your Tool Isn't a Perfect Point**

When we write a CNC program, we often imagine the tool as if it had a sharp, precise point—like drawing with a pencil on paper. In reality, every insert has a rounded tip called the **tool nose radius** (TNR).

Common tool nose radius sizes:

- **0.2 mm** – Ultra-fine finishing operations, tight tolerances
- **0.4 mm** – General finishing work, good surface quality
- **0.8 mm** – Standard roughing and semi-finishing (most common)
- **1.2 mm** – Heavy roughing, maximum strength
- **1.6 mm** – Very heavy roughing in demanding materials

Why do inserts have a radius at all?

1. **Strength** – A sharp point would chip or break immediately under cutting forces
2. **Heat distribution** – The radius spreads heat over a larger area
3. **Surface finish** – A larger radius produces a smoother surface (when used correctly)
4. **Tool life** – Rounded tips last significantly longer than sharp ones

But here's the catch: this same radius that makes the tool practical also makes it geometrically complex.

2.27.2.2 **The Result: Inaccurate Angles and Chamfers**

Without TNRC, the tool's programmed centerline does not match the intended part edge. The controller assumes the tool is a point, but in reality, a circular arc is doing the cutting.

Example 1: The Undersized Chamfer

You program a chamfer:

X40 Z0 → X50 Z-5

Specifications:

- Insert radius: 0.8 mm
- Intended chamfer angle: 45 degrees
- Intended chamfer width: 5 mm on diameter

Actual result without TNRC:

- Chamfer undersized by approximately 1.13 mm on diameter
- Extra material left at both the start and end points
- Part fails inspection

Why does this happen?

When the tool moves along your programmed path, the *center* of the nose radius follows that line. But the actual cutting is happening at the *tangent point* of the radius—which is offset by the radius amount. On a 45-degree chamfer with a 0.8 mm radius, this creates an error of roughly 0.8 mm × √2 ≈ 1.13 mm on diameter.

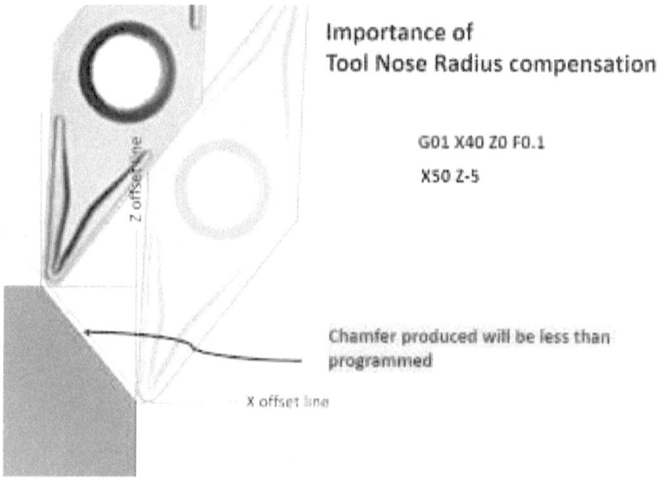

2.27.3 The Bigger Problem: Gouging on Curves

When machining arcs or spherical features, the situation becomes even more critical. The *back edge* of the insert radius can dig into the material if TNRC isn't applied.

Example: Convex Radius (Ball End)

You're machining a 50 mm sphere:

G03 X60 Z-90 R50

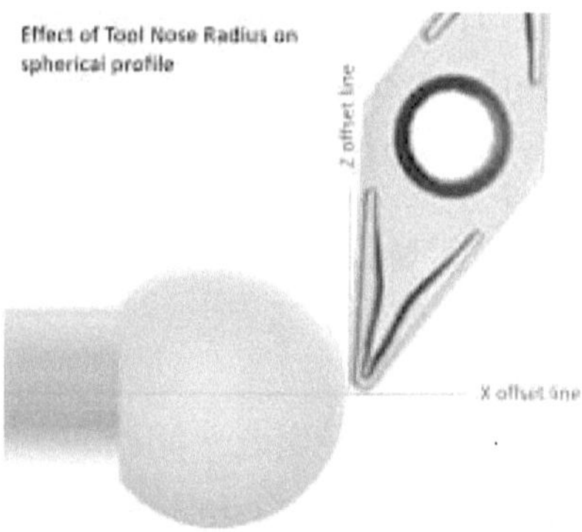

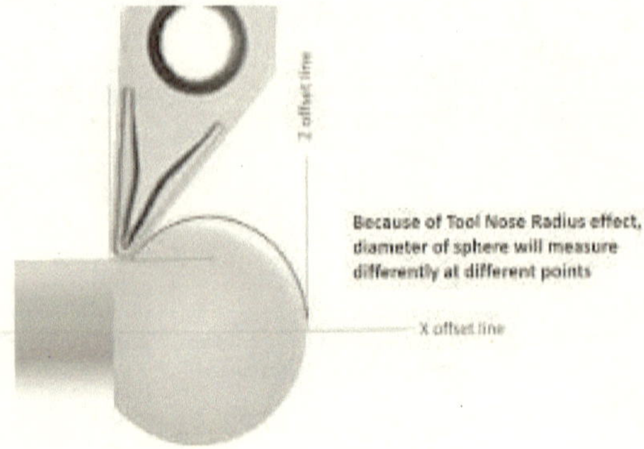

Without TNRC:

- The programmed path is followed by the tool center
- As the tool swings through the arc, the *trailing edge* of the 0.8 mm nose radius interferes with the programmed curve
- Result: Gouges, chatter marks, or a distorted profile that doesn't match your drawing

Takeaway: The bigger the radius and the tighter the geometry, the more critical TNRC becomes. On features like O-ring grooves, ball ends, or complex contours, TNRC isn't optional—it's mandatory for accuracy.

2.27.4 The Solution: How TNRC Works

2.27.4.1 The Smart Adjustment

With TNRC active, the CNC controller automatically shifts the tool's path so that the **cutting edge tangent**

point—not the insert center—follows your programmed geometry.

Think of it like this:

Imagine you're painting a line along a wall with a roller brush. If you measure to the center of the roller, you'll paint in the wrong place. You need to measure to the *edge* where the roller actually touches the wall. TNRC does exactly this for your cutting tool.

The controller's magic:

When you activate G41 or G42, the controller:

1. Reads your programmed tool path
2. Calculates a new, offset path based on the tool nose radius
3. Considers the tool orientation (which way the insert is mounted)
4. Moves the tool center so the cutting tangent follows your original program
5. Continuously updates this offset as direction changes

This ensures that **what you program is exactly what you cut.**

2.27.5 The Key G-Codes

G-Code	Name	Function
G41	Compensation Left	Tool compensates to the left side of the programmed path
G42	Compensation Right	Tool compensates to the right side of the programmed path
G40	Cancel Compensation	Turns off TNRC (must be used before retracting)

Critical Rules:

1. **Activate TNRC during a cutting move** – Use G41/G42 on a G01 line, not during a G00 rapid
2. **Never activate on an arc** – Always turn on TNRC during a straight-line move first
3. **Cancel before rapid moves** – Always use G40 before G00 retraction, or you risk collision
4. **One compensation at a time** – Don't try to use G41 and G42 together

Standard activation sequence:

G00 X42 Z2 (Rapid to approach position)

G41 G01 Z0 F0.2 (Activate compensation during first cutting move)

X52 Z-5 (Continue cutting with compensation active)

G40 G00 X100 Z100 (Cancel, then rapid retract)

2.27.6 G41 vs. G42: Choosing Left or Right

This is where most beginners get confused. But there's a simple, foolproof method.

2.27.6.1 The Golden Rule

"Left or right is decided by looking in the direction of cut."

Imagine you're standing *behind* the tool, riding along with it, looking forward in the direction it's traveling. Ask yourself: "Is the tool on the left or right of the material?"

- If the tool is at **left** of the material → **G41**
- If the tool is at **right** of the material → **G42**

2.27.6.2 Detailed Visual Examples

Example 1 – OD Turning with G42

Path: X40 Z0 → X50 Z-5

Tool direction: Moving towards the chuck (Z0 → Z-5) and increasing diameter (X40 → X50)

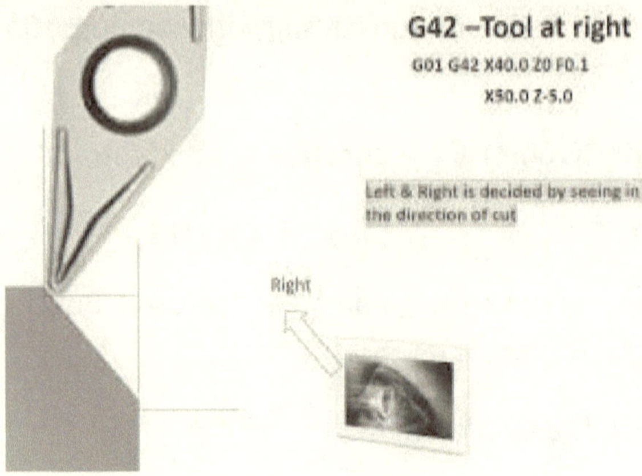

Mental check:

- Stand behind the tool, facing toward Z-
- The workpiece is on your right
- Command: **G42**

Why this works: When turning the outside diameter and moving towards the chuck, you're always cutting with the tool on your right side.

Example 2 – Return Pass with G41

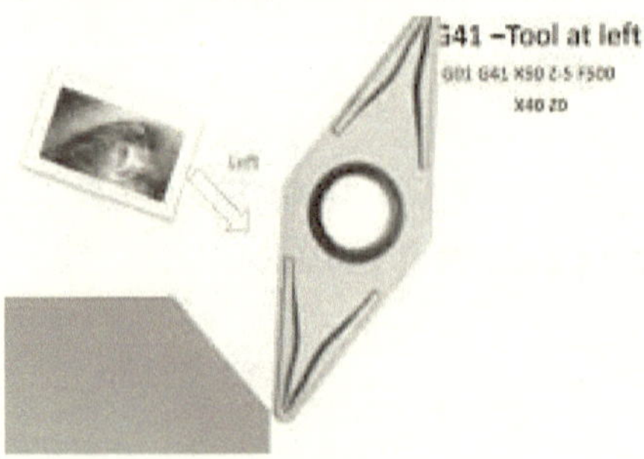

Path: X50 Z-5 → X40 Z0

Tool direction: Moving away from chuck (Z-5 → Z0) and decreasing diameter (X50 → X40)

Mental check:

- Stand behind the tool, facing toward Z+
- The tool is now on your left
- Command: **G41**

Key insight: Same tool, same workpiece, but because the direction reversed, the compensation side reversed too.

2.27.7 Handy Memory Aids

For external turning:

- **OD turning (towards chuck)** → usually **G42**
- **OD turning (away from chuck)** → usually **G41**

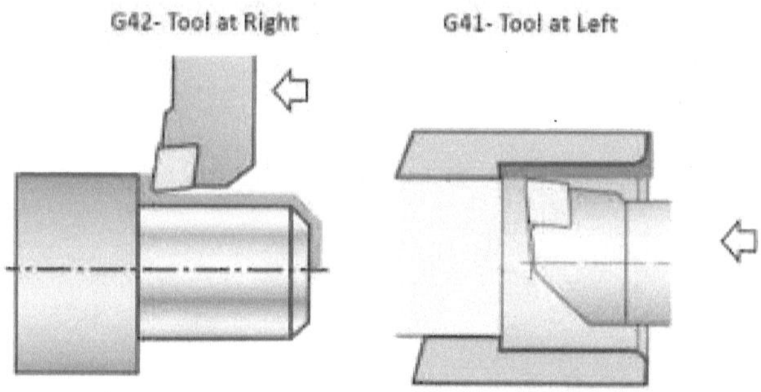

For internal turning:

- **ID boring (towards chuck)** → usually **G41**
- **ID boring (away from chuck)** → usually **G42**

2.27.8 Essential Setup: What the Machine Needs

TNRC won't work unless you feed the CNC two crucial values, usually entered in the **Tool Geometry Offset page** (sometimes called Tool Offset or Tool Table).

2.27.9 Tool Nose Radius (R value)

This is the actual radius of your insert tip, measured in millimeters or inches.

How to find it:

Method 1: Read the insert designation

Insert code: **CNMG120408**

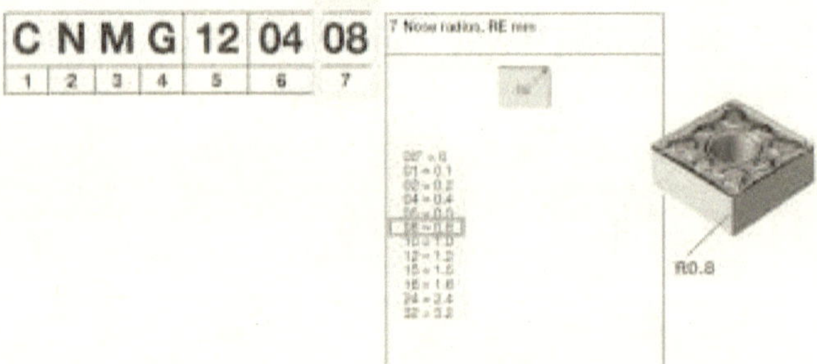

Breaking it down:

- C = 80° diamond shape
- N = 0° clearance angle
- M = tolerance class
- G = hole + chip breaker type

- **12** = Insert size (12.7 mm inscribed circle)
- **04** = Thickness (4.76 mm)
- **08** = Nose radius code → **0.8 mm**

Common nose radius codes:

- 00 = Sharp corner (0 mm) – rarely used
- 01 = 0.1 mm
- 02 = 0.2 mm
- 04 = 0.4 mm
- 08 = 0.8 mm
- 12 = 1.2 mm
- 16 = 1.6 mm

Method 2: Use a radius gauge

Physical radius gauges have curved leaves that match standard radii. Press them against the insert until you find the perfect fit.

Method 3: Check manufacturer's catalog

Every insert has a data sheet specifying the exact nose radius.

Critical: This value must be input accurately. Even a 0.1 mm error will show up on precision parts. If you're unsure, measure it or check the catalog—never guess.

2.27.10 Tool Orientation / Type (Tool Direction Code)

This tells the machine how your insert is physically mounted—which direction it's pointing and which way it cuts. There are **9 standard tool orientation codes** (Types 1–9).

The Four Main Types for Turning:

Type	Description	Common Use	Cutting Direction
3	OD tool, cuts toward chuck	Standard OD turning	Z- direction
4	OD tool, cuts away from chuck	Back-facing, grooving	Z+ direction
2	ID tool, cuts towards chuck	Standard boring	Z- direction
1	ID tool, cuts away chuck	Back-boring	Z+ direction

Visual representation (plan view):

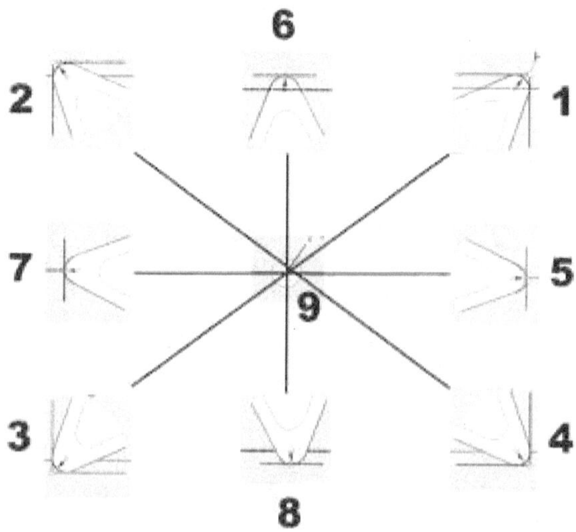

Think of it as a coordinate system:

- **Type 3** = Bottom-left quadrant (most common OD tool)
- **Type 2** = Top-left quadrant
- **Type 4** = Bottom-right quadrant
- **Type 1** = Top-right quadrant

•

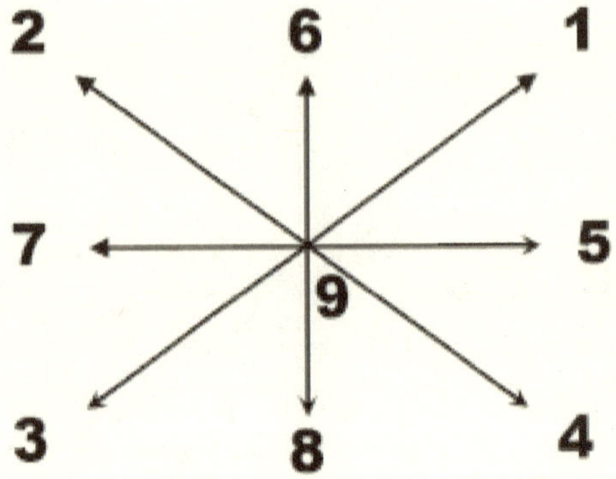

After 4, continue up to 8 and mark 9 at the center

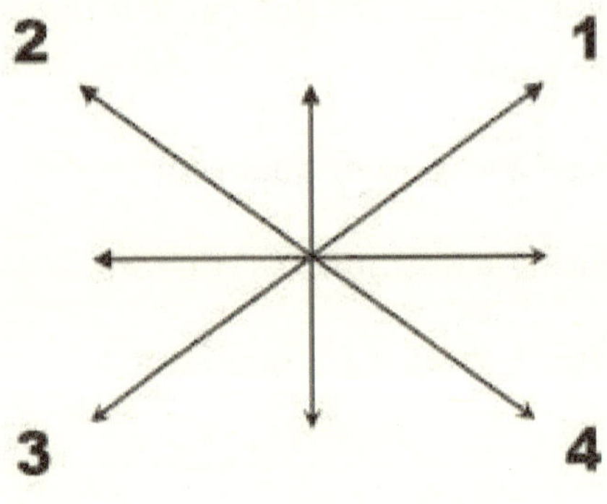

Draw 4 lines as shown and mark types starting from qunadrant1

Types 5-9 are for tools with different orientations

Type1

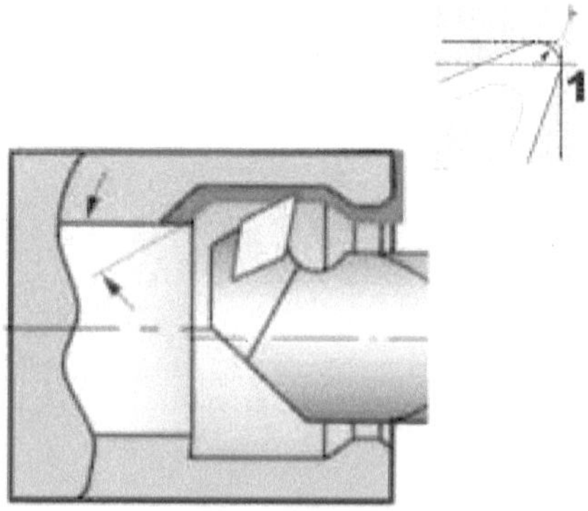

Diameter offset and length offsets are measured on edge of the insert

Type2

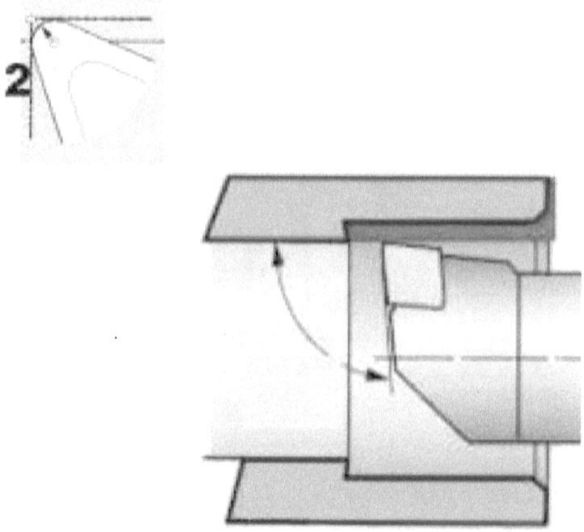

Diameter offset and length offsets are measured on edge of the insert

Type3

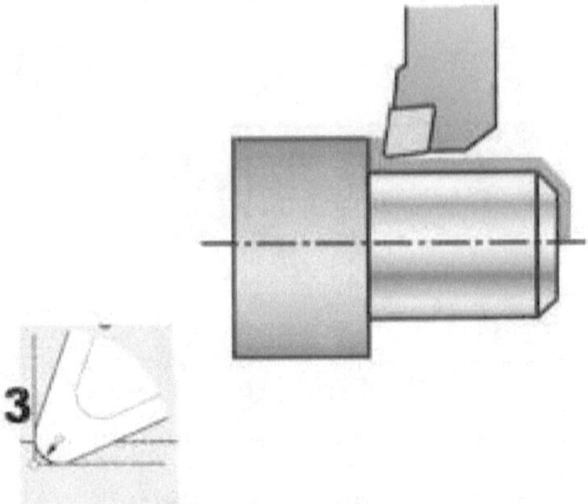

Diameter offset and length offsets are measured on edge of the insert

Type4

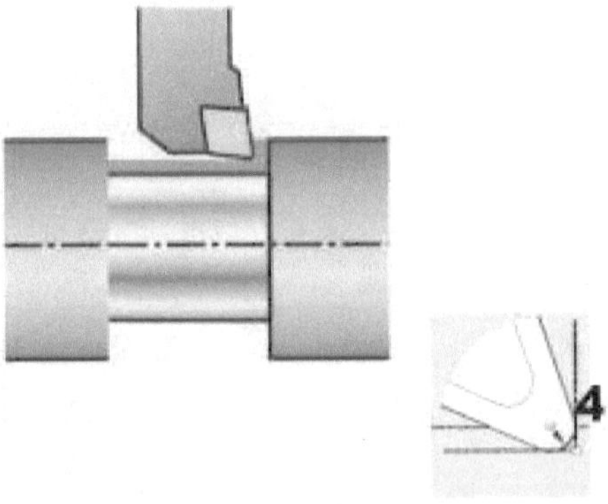

Diameter offset and length offsets are measured on edge of the insert

Type5

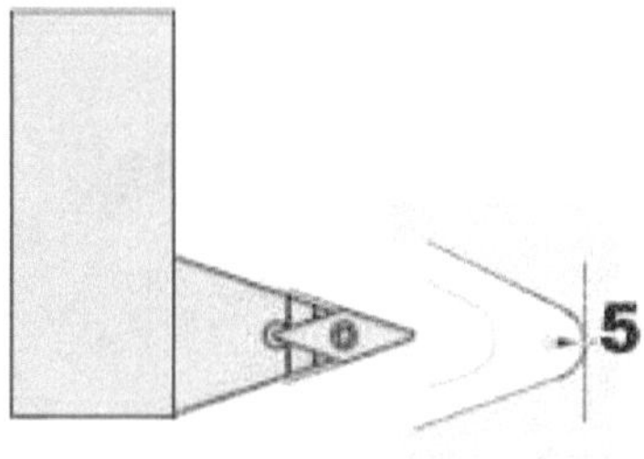

Diameter offset is measured in the center of the insert and length offset is measured on edge of the insert

Type6

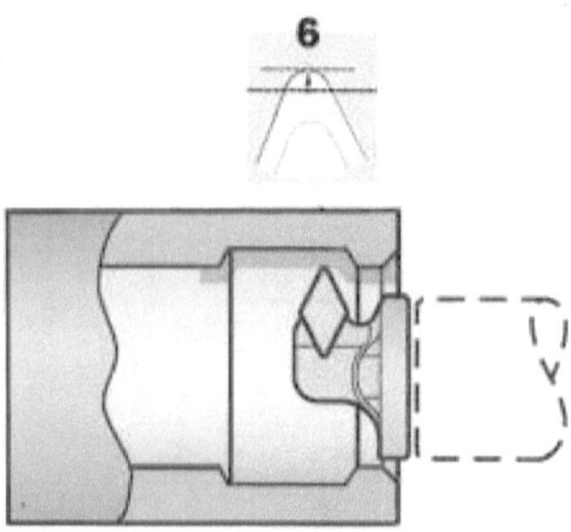

Diameter offset is measured in the edge of the insert and length offset is measured in the center of the insert

Type7

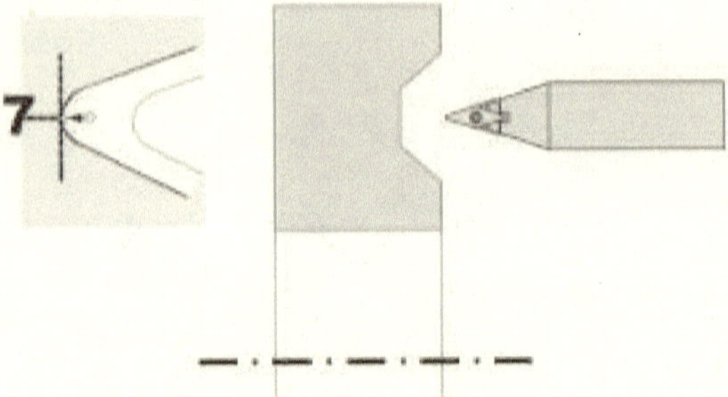

Diameter offset is measured in the center of the insert and length offset is measured on edge of the insert

Type8

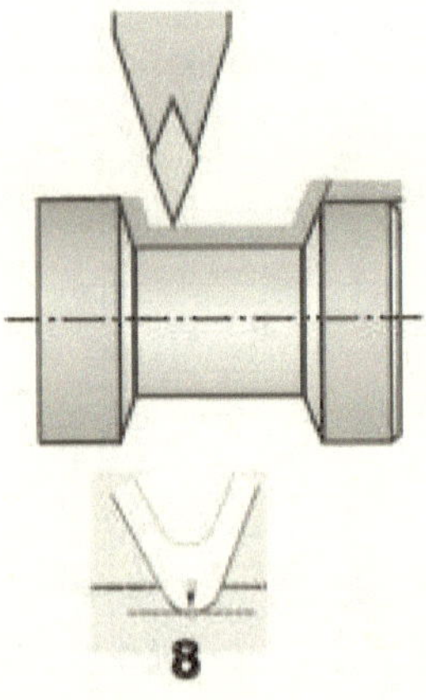

Diameter offset is measured in the edge of the insert and length offset is measured in the center of the insert

Type9

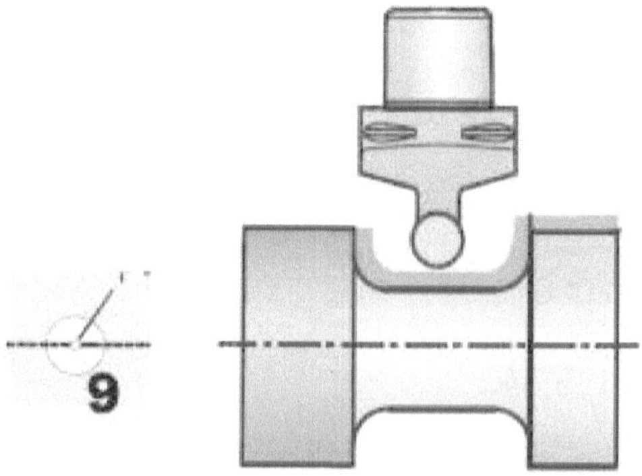

Both diameter offset and length offsets are measured in the center of the insert. Even though the example shown here is for OD, this tool type can be used for any operation like ID or face with a round insert

2.27.11 Entering Tool Data (Example: Fanuc Control)

On most machines:

1. Press **OFFSET** or **OFFSET SETTING** button
2. Navigate to **GEOMETRY** page (sometimes labeled WEAR or TOOL)
3. Select your tool number (e.g., T03)
4. Enter fields:
    - **R** = 0.8 (nose radius)
    - **T** = 3 (tool orientation type)

- **X** = (tool length X offset if needed)
- **Z** = (tool length Z offset)

Example complete tool offset entry:

Tool: T03

X offset: 0.000

Z offset: -125.342

Nose R: 0.800

Type: 3

Without these two values (R and T), TNRC will not work correctly—or may not work at all.

2.27.12 Common Mistakes and Troubleshooting

Mistake: Forgetting G40 Before Rapids

Problem program:

```
G42 G01 X50 Z-20 F0.2
G00 X100 Z100      (Forgot G40!)
```

Result: The tool tries to maintain compensation during the rapid move, potentially crashing into the chuck, tailstock, or workpiece.

Fix: Always cancel first:

```
G42 G01 X50 Z-20 F0.2
G40 G00 X100 Z100    (Correct!)
```

Mistake: Wrong Tool Orientation Type

Symptom: Parts are consistently oversize or undersize by twice the nose radius (e.g., 1.6 mm error with 0.8 mm radius).

Cause: Tool type is set incorrectly (e.g., Type 2 when it should be Type 3).

Fix: Verify the physical tool mounting and set the correct type in the tool table.

Mistake: Activating Compensation on a Rapid Move

Problem:

```
G42 G00 X50 Z-20    (Wrong - G42 with G00)
```

Result: Many controls will alarm out: "G41/G42 not allowed with G00" or similar.

Fix:

```
G00 X52 Z2          (Rapid to approach)
G42 G01 X50 Z-20 F0.2 (Activate on feed move)
```

Mistake: Forgetting to Enter Nose Radius

Symptom: Program runs, but parts are still wrong even with G41/G42.

Cause: Nose radius in tool table is zero or incorrect.

Fix: Check offset page and verify R value matches your insert.

Mistake: Using Wrong G41/G42

Symptom: Parts are mirror-image wrong—oversize where they should be undersize.

Cause: G41 and G42 swapped.

Fix: Use the "look in the direction of cut" rule to verify which compensation side is correct.

2.27.13 Advanced Tips

TNRC and Thread Cutting

Generally, **do not use TNRC for threading**. Threading tools have their own geometry compensation built into the thread cycle (G76 or G92). Using TNRC with threading can cause incorrect thread forms.

TNRC and Grooving

For straight grooving (parting off), TNRC is usually not needed since you're cutting perpendicular. But for **radius grooving** or **form tools**, TNRC can be helpful if your groove has radii or angles. Even in straight grooving, you must compensate the second edge because in grooving both the edge will be cutting.

Very Small Radii (< 0.2 mm)

For extremely fine finishing work with tiny nose radii, TNRC becomes less critical but can still improve accuracy on chamfers and blends.

Check Your Machine Manual

Different controls handle TNRC slightly differently:

- **Fanuc** uses G41/G42 with R and T in the offset table
- **Siemens** uses similar logic but different parameter names
- **Haas** follows Fanuc conventions
- **Mazatrol** (Mazak) may use different terminology

Always consult your specific machine manual for exact syntax.

2.27.14 Conclusion: Your Next Step to Perfect Parts

Tool Nose Radius Compensation (TNRC) bridges the gap between theory and reality in CNC turning. By applying G41/G42 correctly and entering the right tool radius and orientation type, you ensure that the physical insert cuts exactly the profile you program—no manual adjustments, no trial and error, no rework.

Your TNRC checklist:

- ✓ Always look in the direction of cut to decide G41 vs. G42
- ✓ Always enter the correct nose radius in the tool offset table

- ✓ Always enter the correct tool orientation type
- ✓ Always activate compensation during a G01 cutting move, never during G00
- ✓ Always cancel with G40 before rapid moves
- ✓ Always verify your first part with TNRC active—check dimensions carefully

The bottom line: TNRC isn't just a "nice to have" feature. For any part with angles, chamfers, radii, or complex contours, it's the difference between a scrap part and a perfect part. Master this skill, and you'll immediately elevate the quality and consistency of your work.

Next steps:

1. Practice on a simple chamfer with and without TNRC—the contrast will be immediately visible
2. Try a part with a radius and watch how TNRC prevents gouging
3. Experiment with different tool types to understand how orientation affects compensation
4. Check your machine's offset page and verify all your tools have correct R and T values

Welcome to the world of precision CNC machining. With TNRC in your toolkit, you're ready to tackle complex geometries with confidence.

2.28 Direct Angle programming in CNC Turning

2.28.1 Introduction: Why Angles Matter in CNC

Mastering angles is your next step to moving beyond simple straight cuts. It's the key to programming sophisticated features like chamfers and tapers with confidence and precision. This guide will teach you a simple and reliable method for **direct angle programming** using a clock face analogy, making the process intuitive and repeatable.

2.28.2 The Core Concept: The CNC Programming Clock

The foundational mental model for calculating angles in this method is to visualize a standard clock face overlaid on your workpiece drawing. The machine understands angles based on specific reference points on this clock.

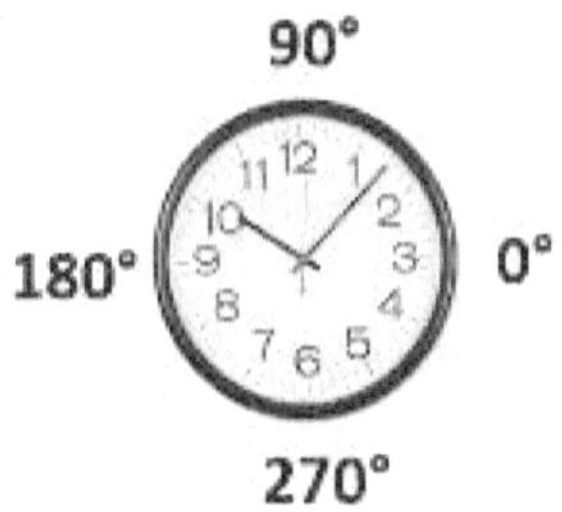

- **0 Degrees (The Starting Point):** Located at the "3 o'clock" position. This corresponds to the horizontal z plus line.
- **90 Degrees:** Located at the "12 o'clock" position.
- **180 Degrees:** Located at the "9 o'clock" position.
- **270 Degrees:** Located at the "6 o'clock" position.
- **Direction of Measurement:** Angles are always measured in a **counter-clockwise** direction, starting from the 0° line at 3 o'clock.

Understanding this clock face is the first step; the next is to use reference lines to apply it to your drawing.

2.28.3 Your Reference Lines: Z Plus and X Plus

Before performing any calculation, the first step is to draw two imaginary lines from the starting point of the angle on your engineering drawing. These lines establish a clear frame of reference.

1. **Horizontal Line (z plus):** This is your primary reference line. It extends horizontally from the angle's start point and corresponds to the 0°/180° axis of the CNC clock.

2. **Vertical Line (x plus):** This is your secondary reference line. It extends vertically from the angle's start point and corresponds to the 90°/270° axis.

These imaginary lines provide the fixed starting points (0°, 90°, 180°, 270°) you need to calculate the final angle value that the machine control can understand.

Now, let's see how to use these reference lines in a few practical examples.

2.28.4 Calculating Angles: Step-by-Step Examples

Example 1: Angle from the Horizontal Line (Upper Quadrant)

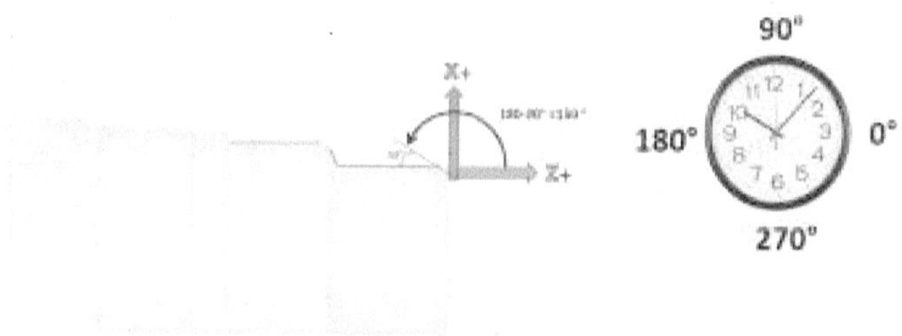

- **The Goal:** Calculate the machine angle for a line specified as 30° from the horizontal.
- **The Reference:** The nearest major reference point is the 9 o'clock position, which is **180°**.
- **The Calculation:** The drawing shows the angle is 30° *back* from the 180° line. Moving back

(clockwise) requires subtraction because it is *against the standard counter-clockwise direction of measurement.* 180 - 30

- **The Final Angle:** 150 degrees.

Example 2: Angle from the Vertical Line (Upper Quadrant)

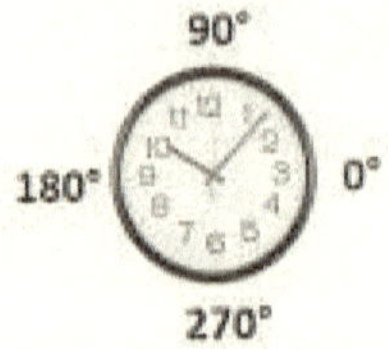

- **The Goal:** Calculate the machine angle for a line specified as 30° from the vertical.

- **The Reference:** The nearest major reference point is the 12 o'clock position, which is **90°**.

- **The Calculation:** This angle is 30° *past* the 90° line, moving *in the standard counter-clockwise direction,* so we add. 90 + 30

- **The Final Angle:** 120 degrees.

Example 3: Angle from the Horizontal Line (Lower Quadrant - ID Turning)

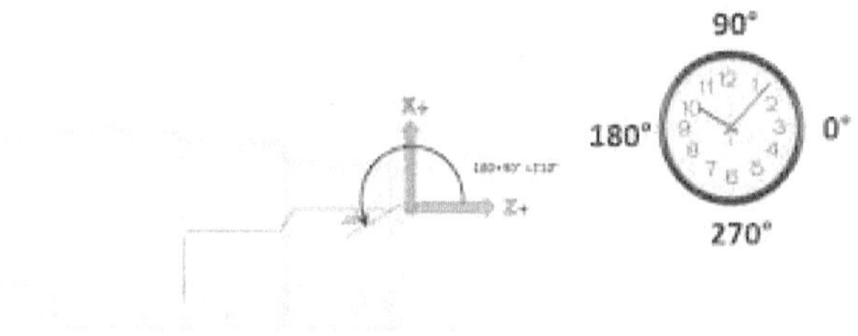

- **The Goal:** Calculate the machine angle for a line specified as 30° from the horizontal in the lower quadrant (common in internal diameter turning, where the tool path is often below the centerline of the part).
- **The Reference:** The nearest major reference point is the 9 o'clock position, which is **180°**.
- **The Calculation:** This angle is 30° *past* the 180° line, continuing *in the standard counter-clockwise direction*, so we add. 180 + 30
- **The Final Angle:** 210 degrees.

Example 4: Angle from the Vertical Line (Lower Quadrant)

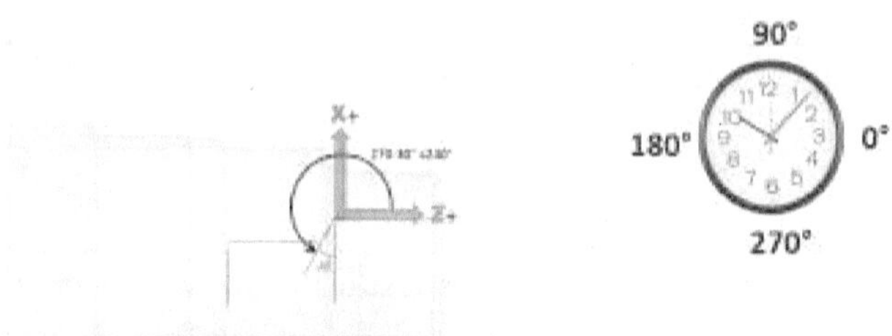

- **The Goal:** Calculate the machine angle for a line specified as 30° from the vertical in the lower quadrant.
- **The Reference:** The nearest major reference point is the 6 o'clock position, which is **270°**.
- **The Calculation:** The angle is 30° *before* the 270° line. Because this requires moving back (clockwise), which is *against the standard direction of measurement*, we subtract. 270 - 30
- **The Final Angle:** 240 degrees.

Once you have confidently calculated the angle, the final step is to format it correctly for your CNC program.

2.28.5 From Calculation to Code: Siemens vs. Fanuc

After calculating the correct angle, you must enter it into the program using the specific syntax for your machine's controller. The format differs between common controllers like Siemens and Fanuc.

Controller	Angle Syntax Example
Siemens	ANG = 150
Fanuc	A150

Pro-Tip: Pairing Angles with Radii In real-world programs, direct angle commands are often paired with "direct rounding" commands in the same block of code to automatically create a radius between features.

- **Siemens:** RND = 4 (for a 4mm radius)
- **Fanuc:** R4 (for a 4mm radius)

2.28.6 Special Cases: 90 and 180 Degree Angles

You can explicitly program ANG = 90 for a vertical toolpath or ANG = 180 for a horizontal toolpath.

However, for simple straight cuts, using these commands is **redundant**. The controller will create a straight vertical or horizontal line by default, so including the angle command doesn't change the outcome. Their real utility is in advanced situations, such as defining "double angles," which makes their inclusion in a simple profile unnecessary.

2.28.7 Conclusion: Your Foundation for Angles

By consistently applying the clock method, you can eliminate guesswork and program angles with confidence. The entire process can be summarized in three simple steps.

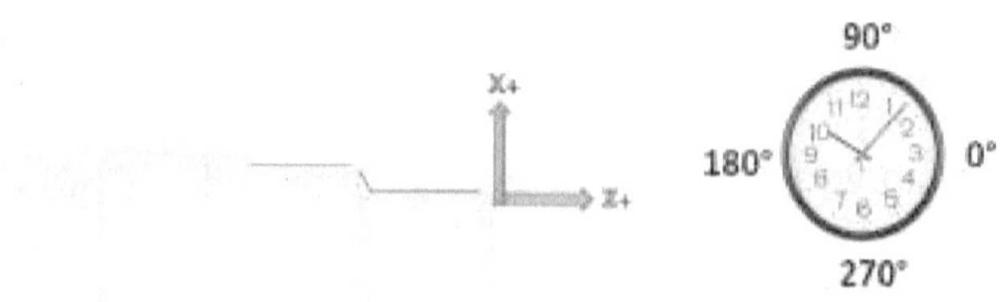

- **Visualize:** Always visualize the CNC clock on your drawing, with 0° at the 3 o'clock position.
- **Reference:** Draw your imaginary horizontal (z plus) and vertical (x plus) lines from the angle's start point.
- **Calculate:** Determine the final angle by adding or subtracting from the nearest 90°, 180°, or 270° reference point.

With this clock method, you now have a reliable mental toolkit to translate any angle from a drawing into machine-ready code, eliminating guesswork and ensuring accuracy every time.

2.28.8 Example

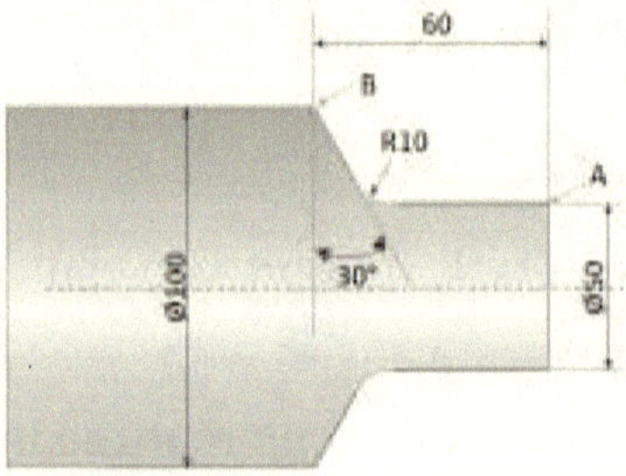


```
N10 G00 X50 Z2
      G01 A180.0 R10.0 F0.15
      G01 X100.0 Z-60.0 A120.0
      Z-80.0
N20 X102.0
```

6.1 Explanation

Entire program is between point A and point B.

N10 G00 X50.0 Z2.0 (Positioning command in Ø50)

G01 A180. R10. F0.15 (A180 is straight line going towards the chuck.

R10 is automatic insertion of radius. This cannot be a stand-alone line. This works along with the next line. Controller understands where to insert the radius based on the next movement)

G01 X100.0 Z-60.0 A120.

What controller does is to draw a line at angle 120°, draw another line at 180° from previous command, take the intersection point and insert radius of R10

2.29 Turning Cycles

2.29.1 Roughing and Finishing

Cycle	Operation	Strategy	Key Parameters
G70	Finishing	Executes stored profile	Uses P–Q blocks from roughing cycle
G71	OD/ID roughing	Axial passes (Z priority)	U: depth/pass, R: retract, U/W: finish allowances
G72	Facing rough	Radial passes (X priority)	Similar to G71 but face-oriented

Cycle	Operation	Strategy	Key Parameters
G73	Pattern roughing	Repeats contour	K: passes/material distribution

2.29.2 Stock removal in turning G71:

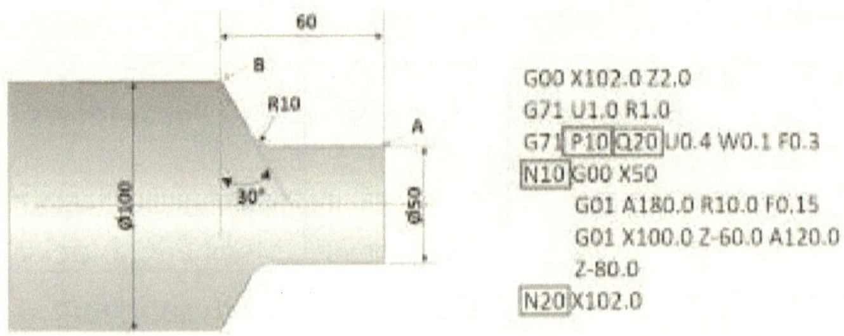

Explanation:

<u>Cycle First line</u>

U1.0 – Depth of cut for roughing

R1.0 – Retract after each cut

<u>Cycle Second line</u>

P10 – Block number of start point

Q20 – Block number of end point

U0.4 - Finish allowance in X

W0.1 - Finish allowance in Z

F0.3 – Feed rate for roughing

2.29.3 Stock removal in Facing G72:

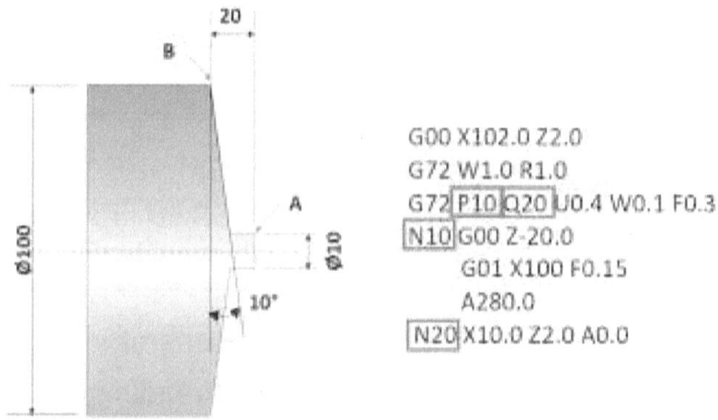

Explanation:

<u>Cycle First line</u>

W1.0 – Depth of cut for roughing

R1.0 – Retract after each cut

<u>Cycle Second line</u>

P10 – Block number of start point

Q20 – Block number of end point

U0.4 - Finish allowance in X

W0.1 - Finish allowance in Z

2.29.4 Grooving and Threading

Cycle	Direction	Motion	Typical Use
G74	Z-axis	Peck feed + retract	End-face grooving, drilling on lathes
G75	X-axis	Peck feed radially	Part-off and OD/ID grooves
G76	X + Z	Multi-pass	General threading
G92	X + Z	Single-pass	Simple or repair threads

2.29.5 End face drilling / grooving cycle G74

Format

```
G74 R_ ;

G74 X(U)_ Z(W)_ P_ Q_ R_ F_ ;
```

Example for peck drilling on end face G74

(Ø10 drill to Z-25, peck 1.5 mm, lift-off 0.3 mm)

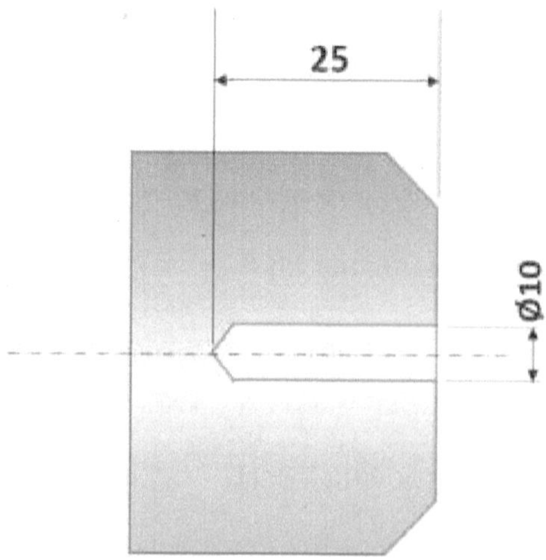

G21 G40 G80 G97

G95 (feed per rev)

T0505 (Ø10 mm drill)

G50 S2500

G96 S120 M03 (CSS on; set for your material)

G00 X0.0 Z2.0 (on center, 2 mm in front of the face)

G74 R0.3 (lift-off in Z after each peck = 0.3 mm)

G74 Z-25.0 Q1500 F0.12 (peck drilling to Z-25; peck = 1.500 mm; feed = 0.12 mm/rev)

G00 Z5.0

M09

G28 U0 W0

M30

Example for peck grooving on face

(Face groove on the end face: 5 mm wide, 5 mm deep at Ø80)

(Tool: 2 mm face-grooving insert)

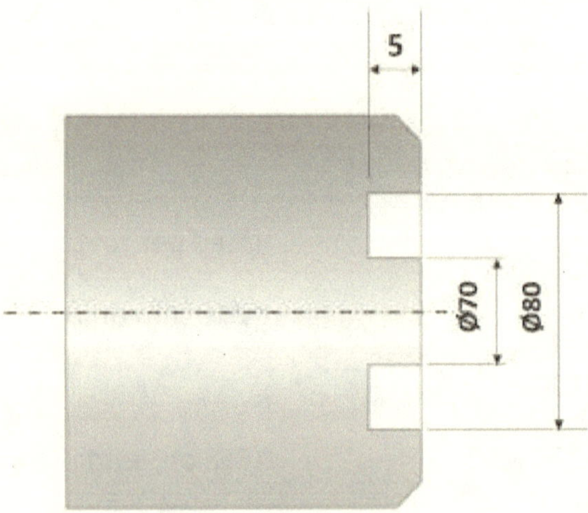

G21 G40 G80 G18 G97

G95

T0303

G50 S2500

G96 S120 M03 (CSS on, adjust to material)

G00 X80.0 Z1.0 (start above face at Ø80, 1 mm in front of Z0)

G74 R0.2 (lift-off in Z after each peck = 0.2 mm)

(From current X (Ø80) step inward to X74 (70+4, tool width compensation)in X-steps of 1.5 mm)

(pecking Z in 0.5 mm bites to Z-5.0, retracting 0.2 mm each peck)

G74 X74.0 Z-5.0 P1500 Q500 R200 F0.08

G00 Z5.0

M09

G28 U0 W0

M30

Note: Values of P, Q & R in line 2 of the cycle are in microns

Example for OD/ID grooving cycle G75

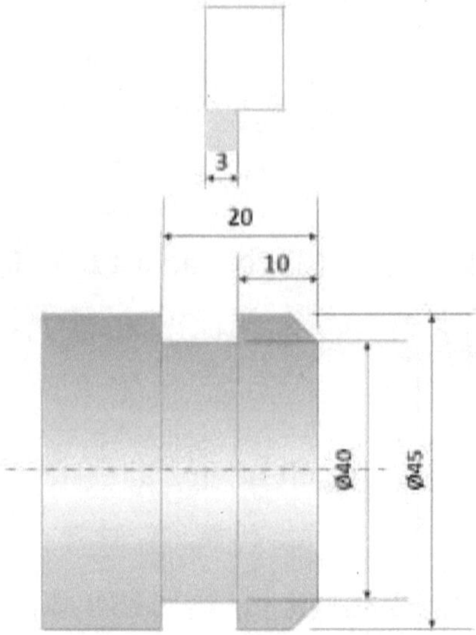

(OD GROOVE: 10.0 W WIDE @ Z-20, FIN DIA Ø40.0)

G21 G40 G80 G97 G99

T0101 (Grooving tool W=3mm)

S1200 M03 (Spindle CW, 1200 rpm)

G00 X47.0 Z2.0 (Safe rapid approach: clear in X)

Z-13.0 (Groove start position, compensated tool width)

G75 R0.05 (Retract 0.05 mm after each peck to break chips)

G75 X40.0 Z-20.0 P2000 Q300 R0.10 F0.08

(X = finished groove DIA (OD) -> Ø40.0

Z = groove end position -> Z-20.0

P = Shift amount in Z after first position -> 2.0 mm -> P2000 (value in microns)

Q = peck DEPTH -> 0.30 mm per peck -> Q300

R = radial FINISH ALLOWANCE -> 0.10 mm left for finish -> R0.10

F = feed per rev -> 0.08 mm/rev)

G00 X50.0

G00 Z2.0 (Retract clear)

M05

G28 U0 W0 (Home)

M30

2.29.6 Screw Thread Machining

When it comes to creating screw threads, that language is built on a specific set of terms. Understanding this vocabulary is the foundational first step for any aspiring machinist. This section will break down the essential terminology for screw threads into simple, easy-to-understand definitions, giving you the confidence you need to move forward.

1. The Core Anatomy of a Screw Thread

Think of a screw thread as a tiny, engineered landscape with its own unique geography. Every thread has several key geometric features, from its highest "peaks" to its deepest "valleys," that define its size and shape.

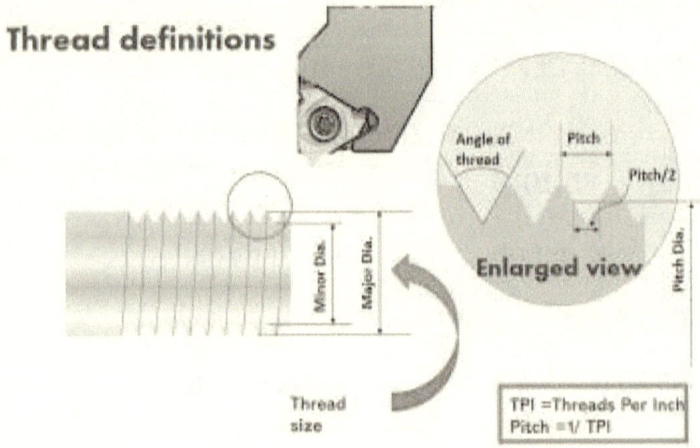

1.1. The Diameters: Measuring a Thread's Size

The primary measurements of any screw thread are its diameters. These three critical dimensions define the overall size and fit of the thread.

1. **Major Diameter:** This is the largest diameter of the screw thread.
2. **Minor Diameter:** This is the smallest diameter of the screw thread.

3. **Pitch Diameter (or Effective Diameter):** This is a theoretical diameter that exists approximately halfway between the Major and Minor diameters. It is a critical value that must be considered when calculating thread height.

1.2. The Thread Profile: Crest, Root, and Flank

The actual shape of an individual thread is defined by three main surfaces that create its profile.

- **Crest:** The top surface of the thread.
- **Root:** The bottom surface of the thread.
- **Flank:** The straight surface that connects the Crest and the Root.

1.3. The Spacing and Angle: Pitch and Thread Angle

Finally, the spacing and angle of the threads are critical measurements that ensure they fit correctly with a mating part.

- **Pitch:** This is the distance between two adjacent crests. For inch-based systems, this concept is often expressed as "Threads Per Inch" (TPI). The pitch can be calculated with the formula: Pitch = 1 / Threads Per Inch.
- **Thread Angle:** This is the angle formed between the two flanks of a thread. A very common angle for ISO Metric threads, for example, is 60 degrees.

Together, these individual parts—the diameters, profile surfaces, pitch, and angle—define the complete geometry of any screw thread.

2. Quick Reference: Key Thread Terms at a Glance

This quick summary table will help you review and reinforce the most important terms from the previous section.

Term	Simple Definition
Major Diameter	The largest diameter of the thread.
Minor Diameter	The smallest diameter of the thread.
Pitch Diameter	The effective diameter, halfway between Major and Minor.
Crest	The very top surface of a single thread.
Root	The bottom surface of the thread.
Flank	The straight side of the thread connecting the Crest and Root.
Pitch	The distance from the crest of one thread to the next.
Thread Angle	The angle between the flanks of the thread.

With these anatomical terms established, we can now look at another fundamental property: the direction of the thread.

3. A Brief Look at Thread Direction: Right-Hand vs. Left-Hand

In addition to their physical dimensions, threads also have a direction, which determines how they tighten and loosen.

- **Right-Hand Thread:** This is the most common type of thread. It is designed to tighten when turned clockwise (to the right).
- **Left-Hand Thread:** This is a less common, special-purpose thread. It is designed to tighten when turned counter-clockwise (to the left).

On a CNC machine, the direction of a thread is determined by the combination of the machine's spindle rotation and the direction the cutting tool travels along the workpiece. A deeper understanding of that process will come later in your training.

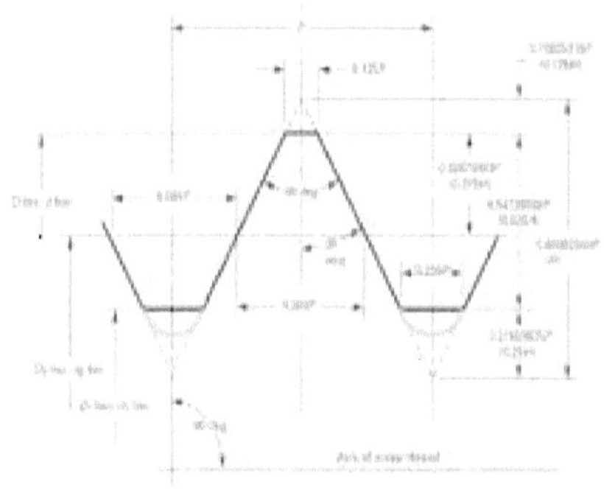

4. Thread height

Above figure is thread profile for UN thread. Fundamental theoretical thread height and actual thread height will vary because of the truncation of threads and root radius.

Thread Form	Included Angle	Formula for Thread Height (h)	Example (P = 1.5 mm)
ISO Metric / Unified	60°	h = 0.61343 × P	0.92 mm
Acme	29°	h = 0.5 × P	0.75 mm
Whitworth	55°	h = 0.64033 × P	0.96 mm
Trapezoidal	30°	h = 0.5 × P	0.75 mm

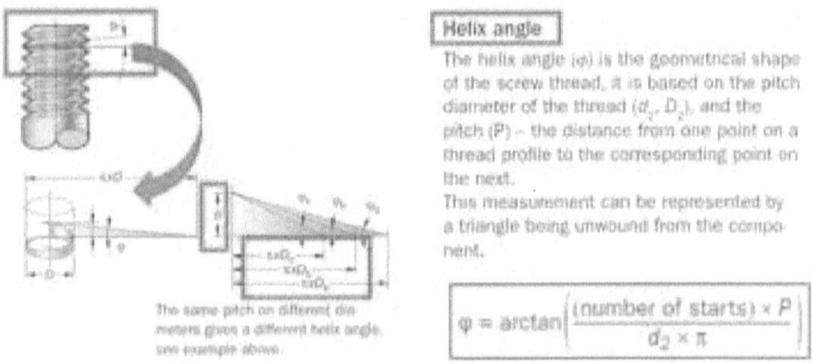

2.29.7 Thread cutting cycle G76 :

G76 P(m)(r)(a) Q(△dmin) R(d) ;

G74X(U)_ Z(W)_ R(i) P(k) Q(△d) F_

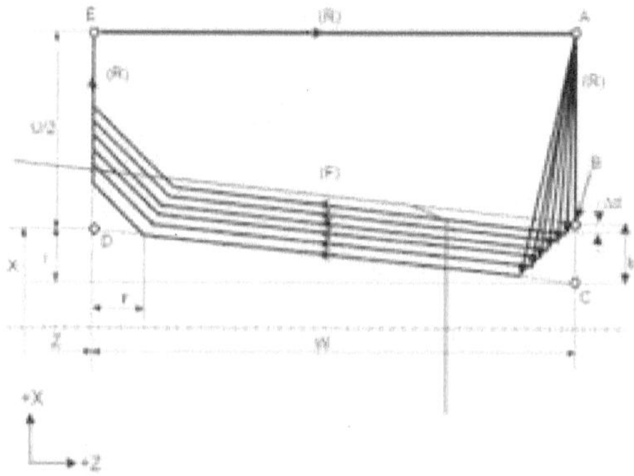

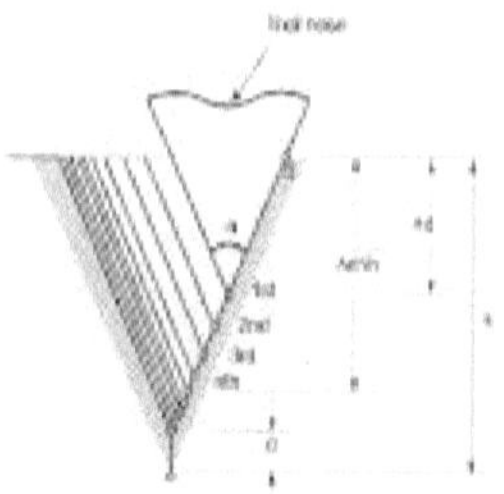

<u>First line</u>

P; first 2 digits m = finish passes, next 2 digits r = chamfer amount, last 2 digits a = tool angle

Q= minimum cutting depth

R= finishing allowance

2nd line

P = thread height k

Q = first-pass depth $\triangle$d

F = pitch/lead.

Program example for **M60 x 2P thread;**

G0 G80 X80.0 Z2.0;

G76 P010160 Q50 R100;

G76 X58.78 Z-50.0 P1220 Q200 F2.0;

G0 G80 X100.0 Z20.0;

Explanation:

First line

P010160 ; 01- 1 finish pass, 0.1 x 2mm (pitch) =0.2, 60 =60° thread angle

Q50 -0.05mm minimum depth of cut for threading.

R100 – 0.1mm finishing allowance

2nd line

X58.78 – Minor diameter of the thread (final diameter)

Z-50.0 – Thread end point Z-50.0

P1220 – Total thread depth (0.61343 × P =1.22)

Q200 – First cut depth =0.2mm

2.29.8 G71 / G70 Programming Example

```
%
O2001
G50 S2000
T0101 M08
G96 S180 M03
G00 X82. Z5.
G71 U2.5 R0.5
G71 P10 Q20 U0.4 W0.1 F0.3
N10 G00 X30.
G01 Z0 F0.2
X40. Z-5.
Z-30.
X60. Z-55.
Z-80.
N20 X80.
G70 P10 Q20
M30
```

2.30 Turn-mill center and C-Axis Machining

2.30.1 Introduction: What is a Turn-Mill Center?

A turn-mill center with a C-axis is fundamentally a three-axis machine, combining the linear movements of the **X-axis** and **Z-axis** with the rotational movement of the **C-axis**. A turn-mill center is basically used for light milling /drilling applications using a live tool

This section will focus on the foundational concepts behind C-axis milling and drilling operations, breaking down how these machines create complex features on a workpiece. To begin, let's explore the core movements that make these operations possible: the coordinated action of the C-axis and the X-axis.

2.30.2 The Core Movements: How the X-Axis and C-Axis Work Together

In C-axis operations, two primary movements are coordinated to cut features into a part. Understanding how they interact is the key to understanding the machine's capabilities and limitations.

- **The X-Axis:** This axis controls the linear, in-and-out movement of the cutting tool. Its job is to move the tool *towards* and *away* from the central axis of the workpiece.

- **The C-Axis:** This is the controlled *rotation* of the workpiece. It can either rotate to a specific angle and hold (indexing) or rotate continuously to coordinate with the X-axis.

The most critical limitation of a basic turn-mill center is the absence of a Y-axis. This means the cutting tool or drill cannot move up and down on its own. Therefore, all shapes and features are created by precisely coordinating the in-and-out motion of the **X-axis** with the precise rotation of the **C-axis**.

Now, let's see how these coordinated movements are applied in practical milling and drilling examples.

2.30.3 C-Axis in Action: Milling Operations

Milling is a machining process that uses a rotating multi-point cutting tool to remove material from a workpiece. On a turn-mill center, this is achieved by combining the tool's rotation with the X-axis and C-axis movements.

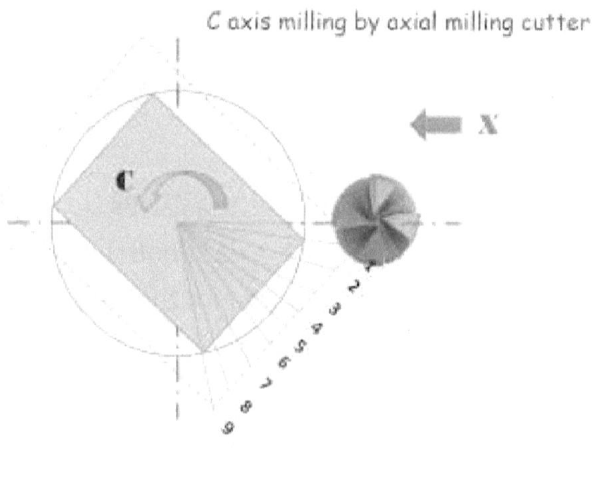

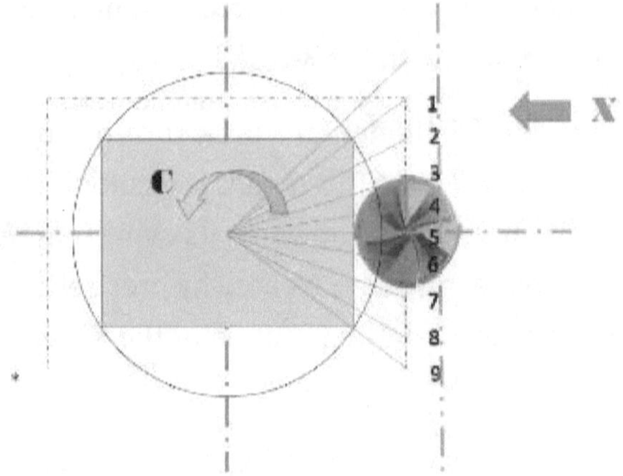

2.30.4 Example 1: Creating a Rectangular Block with an Axial Cutter

A common demonstration of C-axis capability is milling a round bar into a rectangular block. To create a flat side on a round part, the machine must perform a coordinated move.

1. **Cutter Rotation:** The axial milling cutter spins in a fixed position.

2. **Coordinated Motion:** As the **C-axis** slowly rotates the part, the **X-axis** simultaneously moves the cutter toward and then away from the center of the part.

3. **Generating a Flat Surface:** This simultaneous, dynamic motion of both axes is what generates a true flat surface, distinguishing it from a simple plunge cut where only one axis moves.
4. **Repetition:** This coordinated move is repeated for each of the four sides to form the complete rectangular shape.

2.30.5 Example 2: The Challenge of Milling with a Radial Cutter

Another approach involves using a radial milling cutter, where the tool cuts with its side rather than its face. However, this method introduces a significant challenge.

The core problem with this method is "**fouling.**" This occurs when the curve of the radial cutter accidentally removes extra material from an adjacent part of the workpiece. As the cutter moves to a specific coordinate, its circular shape can gouge areas it isn't supposed to, as illustrated by the red areas in the diagram.

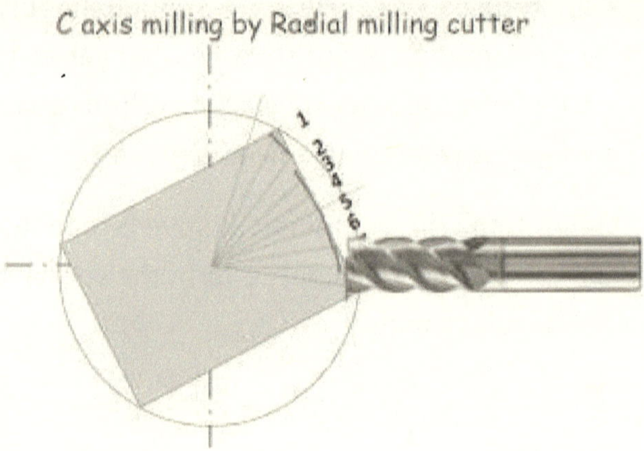

Fortunately, there are two primary solutions to overcome this limitation:

1. **Use a Larger End Mill**

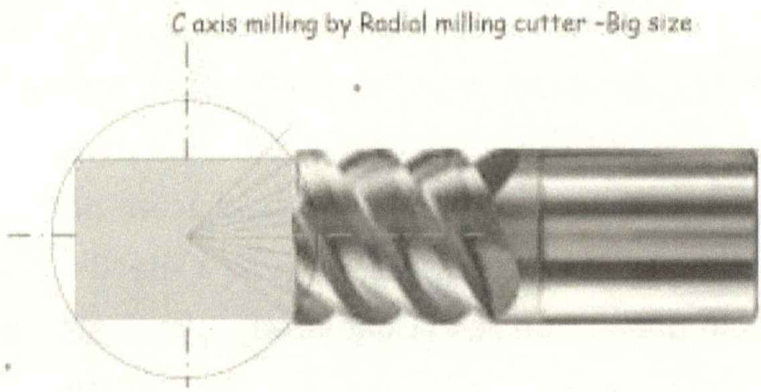

If the diameter of the end mill is bigger than the entire area being milled, it can machine the surface in a single pass without causing fouling on adjacent surfaces.

2. **Use a Ball Endmill**

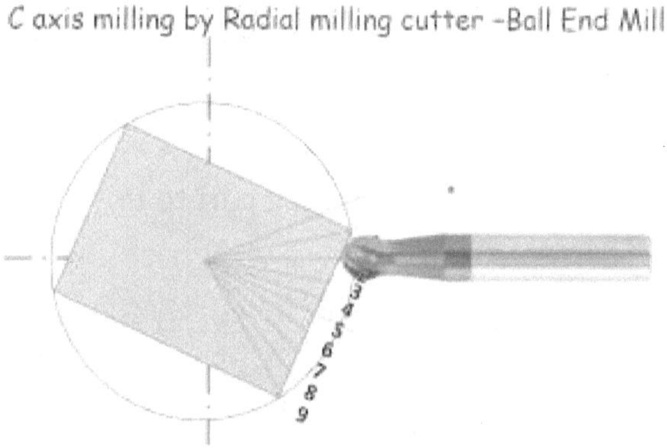

A ball endmill, with its spherical tip, can avoid fouling by making small, precise movements that resemble 3D motion. This creates a toolpath that mimics a 3D carving motion, allowing it to machine the flat surface without the tool's side fouling the adjacent material. This advanced technique often requires specialized software to program the complex toolpaths correctly.

While milling presents certain challenges, drilling operations reveal a more fundamental limitation of the basic C-axis machine.

2.30.6 C-Axis in Action: The Limitation of Radial Drilling

Imagine the goal is to drill a hole where the hole's central axis is offset from the central axis of the part. This is a common requirement in manufacturing.

On a C-axis turn-mill center, the machine can use its two axes to align the drill with the desired *position* for the hole:

- The **X-axis** moves the drill out from the center.
- The **C-axis** rotates the part to the correct angle.

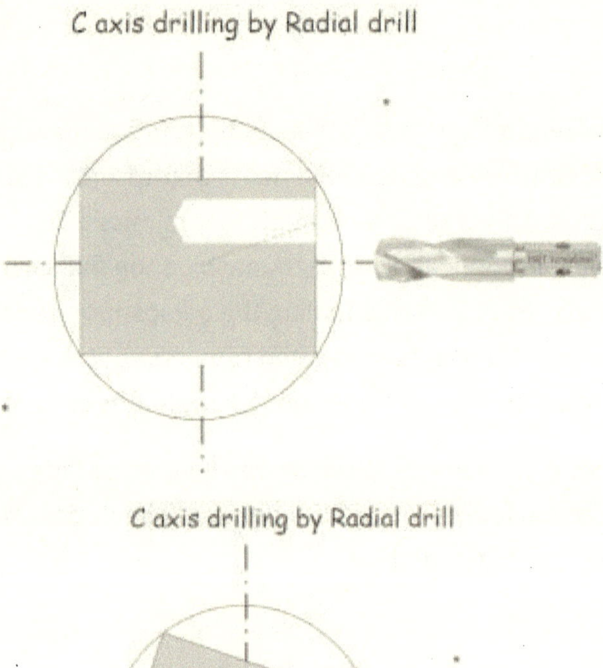

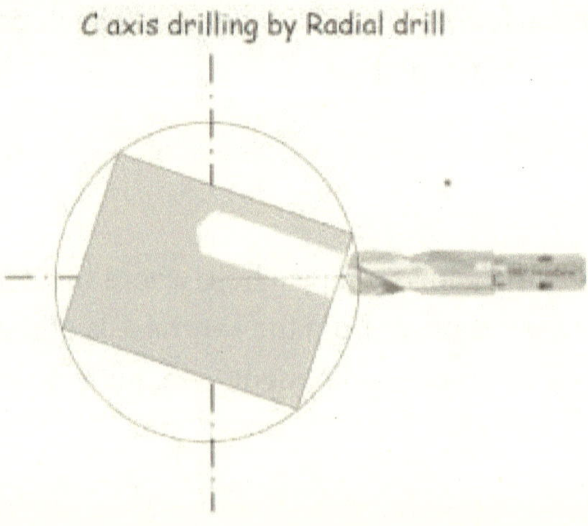

However, the operation is ultimately not possible. The critical limitation is that the drill itself cannot move up or down (no Y-axis). It can only drill on an axis that is *parallel* to the part's central axis. Even though the drill is positioned correctly, its path of travel is fixed and always aligned with the center.

This illustrates a core principle of machining: you must always account for the machine's fundamental axes of movement. On a basic turn-mill center, the absence of a Y-axis makes drilling a truly offset hole impossible.

Understanding these capabilities and, more importantly, these limitations is crucial for effective machining. Let us summarize the key principles to remember.

2.30.7 Key Takeaways

As you begin your journey into CNC machining, remember these three fundamental concepts about C-axis turn-mill centers.

- **Coordinated Motion is Key:** The C-axis provides rotation for the workpiece, which works in tandem with the X-axis's in-and-out motion to create shapes and features.
- **Milling Methods Have Trade-offs:** While axial milling is straightforward, radial milling can cause "fouling" (removing extra material), which requires specific solutions like using a different type of tool.

- **Know Your Machine's Limits:** The absence of a Y-axis is a fundamental constraint that makes certain operations, like drilling truly offset holes, impossible on a basic C-axis turn-mill center.
- **Solution:** A turn-mill center with Y-axis (will become 4axis machine), can get rid of all these limitations.

2.30.8 Face milling using C-axis G12.1/G112

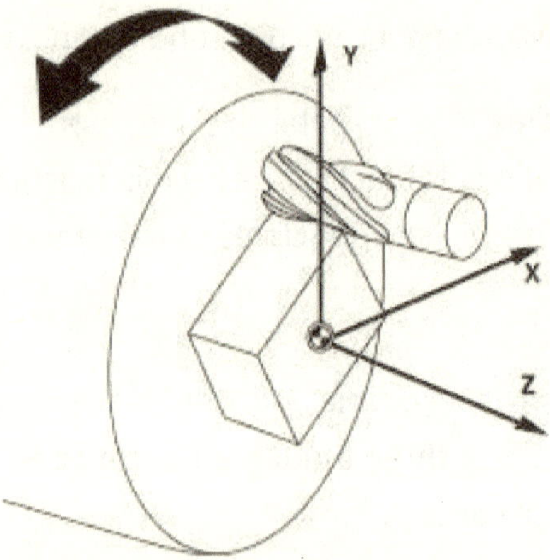

6.1 Face milling EXAMPLE

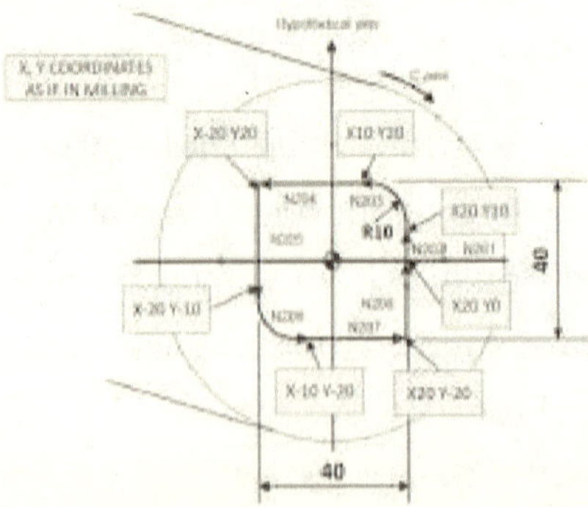

Calculate the coordinates as shown in the figure.

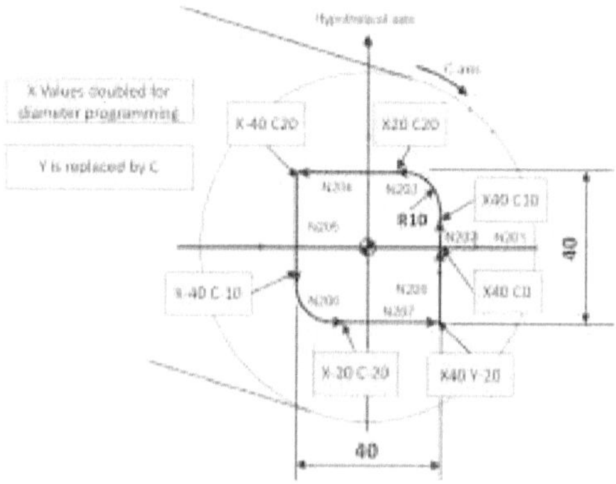

Double the values of X for diameter programming.
Replace the Y coordinate with C, as shown in the figure.

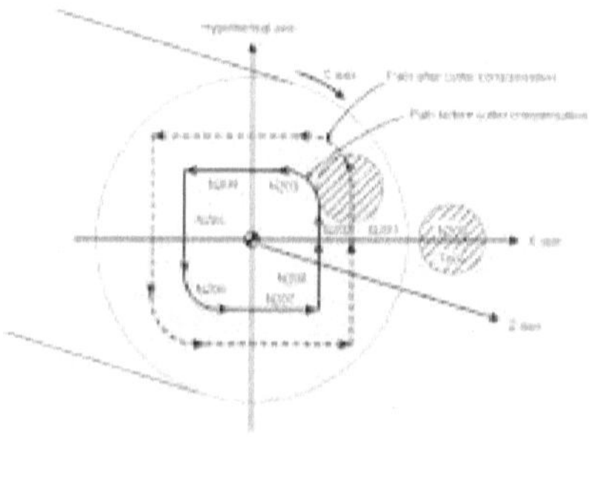

O0001;

;

N010 T0101

;

N0100 G90 G00 X60.0 C0 Z-1.0 ; Positioning to start point

N0200 **G12.1** ; Start of polar coordinate interpolation

```
N0201 G42 G01 X40.0F100.;
N0202 C10.0;
N0203 G03 X20.0 C20.0 R10.0;
N0204 G01 X-40.0;
N0205 C-10.0;
N0206 G03 X-20.0 C-20.0 I10.0 J0;
N0207 G01 X40.0;
N0208 C0;
N0209 G40 X60.0;
```

(N0201–N0209: Geometry program (program based on cartesian coordinates on X axis–hypothetical axis plane))

N0210 **G13.1** ; Cancellation of polar coordinate interpolation

```
N0300 G0 Z10.0;
N0400 X100.0C0;
.
N0900M30;
```

Note: Check your machine's documentation for choosing between G12.1 & G112 or G13.1 & G113.

2.30.9 Face drilling EXAMPLE

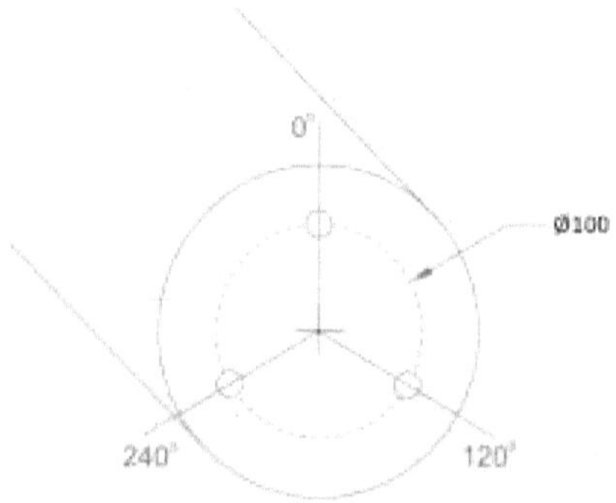

For drilling use simple coordinates using X & C

```
O0100

T0101

G54 G00 X100.0 Z5.0

G98

M154 (C-Axis engage)

C0.0

M133 P2000 (Live Tooling Drive Forward)

G81 Z-10.0 F50.0

C120.0

C240.0

G00 G80 Z5.0

M155 (C-Axis Disengage)
```

M135 (Live tool stop)

G53 X0

G53 Z0

M30

Note: Check your machine's documentation for choosing the right M codes for C-axis engage, live tool rotation etc.

2.30.10 Radial Drill example

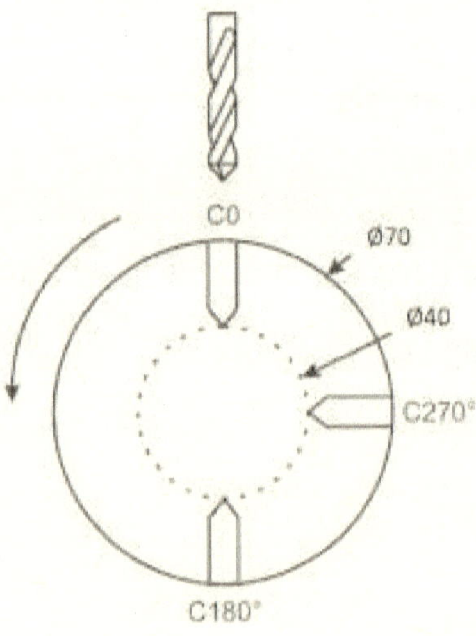

```
(LIVE DRILL - RADIAL)
T101
G19
G98
M154     (Engage C-axis)
G00 G54 X80. C0. Z50.
G00 Z-20.0
G00 X72.0
G97 P1500 M133
M08
G00 C0.
G19 G75 X40. Q1.0 F100.
G00 C180.
G19 G75 X40. Q1.0 F100.
G00 C270.
G19 G75 X40. Q1.0 F100.
G00 G80 Z50. M09
M135
M155
M09
```

```
G00 X150. Z100.

G18

G99

M30

%
```

Note: Check your machine's documentation for choosing the right M codes for C-axis engage, live tool rotation etc.

Disclaimer: Even though details explained in this book will be applicable and accurate for most of the Fanuc controller models, you may find few differences. It is highly recommended to read the specific machine manual to get the information specific to your machine.

2.30.11 Programming Best Practices

- **Start with Safety Block:** Reset modal states with G90 G17 G21 G40 G49 G80 G94.
- **Use Absolute Positioning (G90):** Reduces cumulative positioning errors versus incremental (G91).
- **Handle Tool Changes Carefully:** Retract tool, stop spindle and coolant before changing (M05, M09, M06).

- **Cancel Modal Codes at End:** Use G40, G49, G80 to clear modes and avoid unexpected carryovers.
- **Optimize Feeds and Speeds:** Rough with higher feed rates and moderate speeds; finish with slower feeds and higher rpm for surface quality.
- **Comment Generously:** Include program info for clarity—part number, material, tooling, programmer.
- **Standard Format:** Use block numbers (N10, N20...) and consistent spacing for readability.
- **Limit Incremental Movements:** G91 can cause drift; prefer absolute positioning; verify coordinates regularly.

2.30.12 Program Verification & Testing

Verification Steps

- **Simulation:** Use CAM or verification software to review toolpath.
- **Dry Run:** Execute above part (e.g., Z + 50 mm) with stops enabled.
- **Single Block Execution:** Step through instructions line by line.
- **Validate Offsets:** Confirm work offsets (G54–G59) and tool length offsets (H codes) with probes or edge finders.

2.30.13 Common Issues & Fixes

Issue	Cause	Fix
Unexpected toolpaths	Modal codes left active	Cancel with G40, G49, G80.
Feed rate errors	Missing or incorrect F values	Specify feed in every cutting block.
Arc errors	Wrong I, J, K, or plane selection	Verify arc center and plane G17–G19.
Collisions	Incorrect offsets or retracts	Check G43 H, G54, safe Z moves.

Pro Tip: *Always perform first-article inspection after program verification to ensure dimensional accuracy.*

2.31 Advanced Programming

- **Parametric Programming:** Use variables (#1, #2) for dynamic, adaptable code (e.g., #1 = 10 sets pocket depth).

- **Macros:** Use M98 calls with repeat counts for repetitive features (e.g., M98 P1002 L5).
- **Conditional Logic:** Incorporate IF/THEN for adaptive machining parameters (e.g., adjust feed on spindle load).
- **Custom Cycles:** Tailor canned cycles for specific complex tasks like deep hole drilling.

Refer chapter10 for more details.

Standards & Compatibility

- **ISO 6983:** The international standard ensuring consistent CNC programming language use globally.
- **Best Practice:** Always consult specific machine documentation for supported codes, canned cycles, and macros.

2.32 5-Axis Machining

2.32.1 Introduction

This section serves as a clear starting point for understanding the fundamental concepts of 5-axis CNC machining. A crucial first step on your learning journey is to grasp the basic machine setups—a core concept that even many experienced programmers do not fully understand. This knowledge is critical because the machine's configuration directly impacts how you plan your work, design your fixtures, and write your G-code programs. This section will walk you through the three primary configurations of 5-axis machines, providing the foundational knowledge you need.

These machine classifications are based on a simple but critical distinction: where the machine's extra axes of motion are located.

2.32.2 The Building Blocks: Tilting and Rotary Axes

A 5-axis machine builds upon a standard 3-axis machine (X, Y, and Z) by adding two additional axes of motion. These machines are classified based on where these two axes—the **tilting** and **rotary** axes—are physically placed on the machine.

- **Tilting Axis:** This axis tilts the cutting tool or the workpiece to achieve a specific angle. It is often referred to as the **A-axis** when it tilts around the machine's X-axis, or the **B-axis** when it tilts around the Y-axis.
- **Rotary Axis:** This axis rotates the cutting tool or the workpiece. This is commonly the **C-axis**, which rotates around the machine's primary Z-axis.

The specific arrangement of these tilting and rotary axes on either the machine's main table or its spindle head gives us three distinct machine configurations.

2.32.3 The Three Main 5-Axis Configurations

Depending on whether the machine's table or its head (the part that holds the spindle, which is the rotating component that grips the cutting tool) provides the tilting and rotary motion, we get three different setups. Understanding these configurations is key to understanding how a particular machine will approach a complex part.

Configuration Type	Tilting Axis Location	Rotary Axis Location
Table-Table	Table	Table
Head-Table	Head	Table
Head-Head	Head	Head

2.32.3.1 Table-Table Configuration

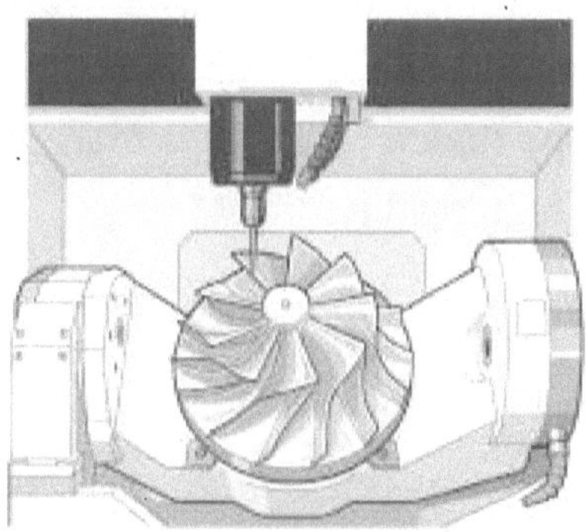

In a **Table-Table** machine, both the tilting axis and the rotary axis are built into the machine's table. For example, a common setup involves an **A-axis** (tilting around the X-axis) and a **C-axis** (rotating around the Z-axis), both integrated into the work-holding table. This setup is excellent for smaller, more complex parts but the movement of the table can limit the maximum weight of the workpiece.

2.32.3.2 Head-Table Configuration

In a **Head-Table** machine, the motion is split. One axis is located on the machine's head, while the other is on its table. A typical example is a machine with a **B-axis** (a tilting axis) configured on the head and a **C-axis** (a rotary axis) built into the table. This hybrid approach offers a versatile balance, often allowing for a larger work envelope than a Table-Table machine while maintaining high rigidity.

2.32.3.3 Head-Head Configuration

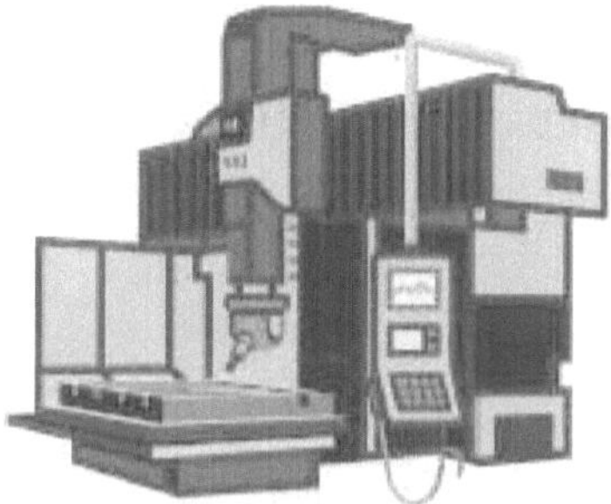

In a **Head-Head** machine, both the tilting and rotary axes are located on the machine's head assembly, which holds the spindle. A common configuration would feature a **B-axis** for tilting and a **C-axis** for rotation, with both motions originating from the head itself. Because the workpiece remains stationary on the table, this configuration is ideal for machining very large and heavy parts that would be difficult to move with precision.

2.32.4 A Critical Concept: Tool Center Point Management (TCPM)

Tool Center Point Management (TCPM) is an essential function in 5-axis machining that makes complex, angled cuts possible and safe. In simple terms, TCPM's job is to ensure that the very tip of the cutting tool remains exactly where it needs to be on the workpiece, even while the machine's head or table is tilting and rotating to a new angle. TCPM dynamically recalculates the tool's position using real-time kinematic transformations, allowing programmers to define toolpaths in part coordinates without recalculating offsets for each orientation change

The difference between machining with and without this function is dramatic.

TCPM OFF

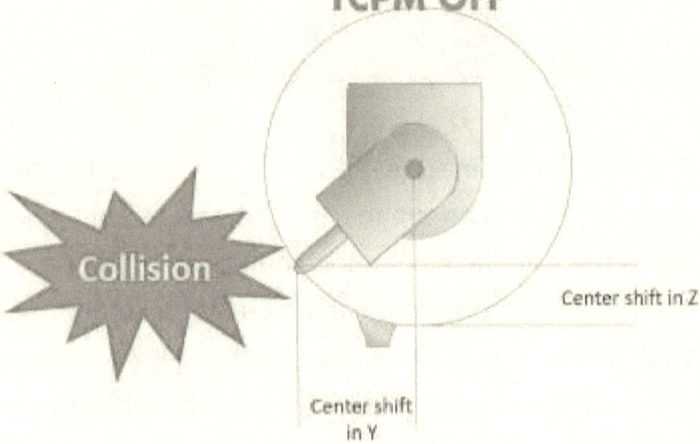

TCPM ON

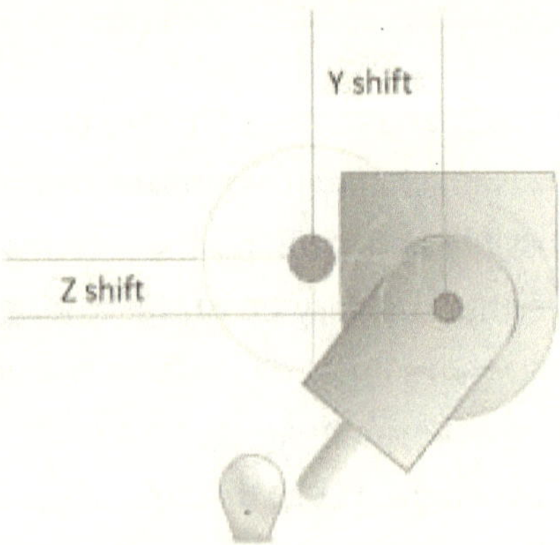

Without TCPM (TCPM OFF)	With TCPM (TCPM ON)
When a command to tilt is given, the machine simply rotates around its fixed mechanical pivot point.	The machine's controller automatically calculates the necessary compensation and moves the linear axes (X, Y, Z) to counteract the effect of the tilt.
This rotation causes the tool tip to swing away from its intended position. This creates an unintended shift in the tool's linear (X, Y, and Z) position, plunging it into the part.	The **tool center point remains perfectly stationary** on the workpiece. The controller accomplishes this by moving the linear axes (X, Y, Z) to shift the machine's physical pivot point, ensuring the tip of the tool doesn't move as the angle is achieved. This proactive compensation is what prevents collisions.

Mastering this distinction between TCPM ON and OFF is often what separates a novice programmer from a professional. It is the core software feature that makes safe and accurate 5-axis simultaneous motion possible.

Practical tip: Even though the controller calculates the dynamic positions, it is advisable to keep the part as close to the center of the table as possible to reduce the mechanical errors.

2.32.5 Tool Center Point Management (TCPM) in Fanuc G43.4

Syntax

G43.4 Hxx

- **Hxx**: Refers to the tool length offset register.
- TCPM remains active until canceled by **G49**.

Key Points

- G43.4 enables **kinematic transformation** for tool tip control.
- It requires a properly calibrated **rotary axis center** and **kinematic model**.
- When active, the control automatically manages complex 3D motion for smooth tool movement.

Example

G90 G17 G21

T5 M6

G43.4 H05 Z100.

G1 ----

...

G49 (Cancel TCPM)

Benefits

- Simplifies 5-axis programming
- Ensures consistent tool tip accuracy during rotary motion
- Reduces the need for complex manual transformations

2.32.6 Coordinate Rotation 2D – G68

The **G68 command** enables the rotation of the active coordinate system in a CNC machine about a specified point. This function allows the programmer to rotate the machining path without physically reorienting the part or redrawing the geometry in the CAM system. It is particularly useful when machining features that are repeated at angular intervals or when aligning machining coordinates with the part's actual orientation on the fixture.

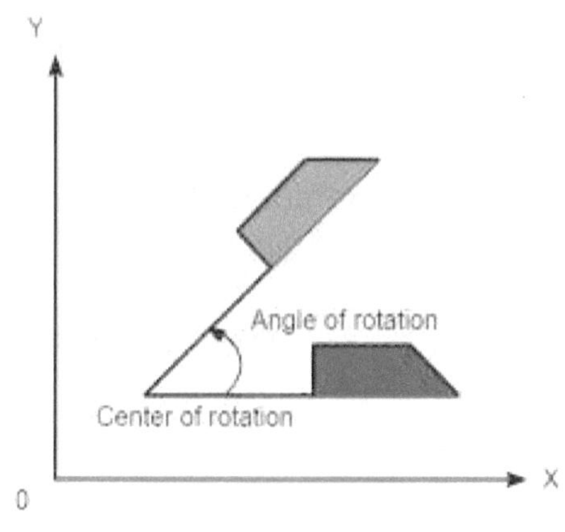

Syntax

G68 X_ Y_ R_

- **X, Y** — Specify the rotation center (pivot point).
- **R** — Specifies the angle of rotation in degrees (positive = counterclockwise, negative = clockwise).

Example

G68 X0. Y0. R45.

This command rotates the coordinate system **45° counterclockwise** about the origin (X0, Y0). All subsequent moves are interpreted relative to the rotated axes until rotation is canceled.

Canceling Rotation

Use **G69** to cancel the coordinate rotation and return to the normal orientation.

Application Notes

- Commonly used in 2D milling or drilling of patterns at an angle.
- Simplifies programming of angular features without trigonometric calculation.
- Rotation is purely a coordinate transformation—machine zero remains unchanged.

2.32.7 Coordinate System Rotation in 3D Space – G68.2

2.32.7.1 Overview

The **G68.2** command in Fanuc and compatible CNC controls is used to **rotate the active coordinate system in three-dimensional space**. Unlike simple 2D plane rotations (such as G68 for XY-plane), G68.2 allows the programmer to define a **tilted working plane** by specifying rotation angles about the machine's X, Y, and Z axes—enabling complex multi-axis machining setups such as 5-axis positioning and 3+2 machining.

This function is essential when machining features on angled surfaces, such as holes on an impeller blade or pockets on an inclined face. By reorienting the coordinate system virtually, programmers can write toolpaths as if the workpiece were aligned flat in the standard plane, simplifying programming and reducing the need for trigonometric calculation.

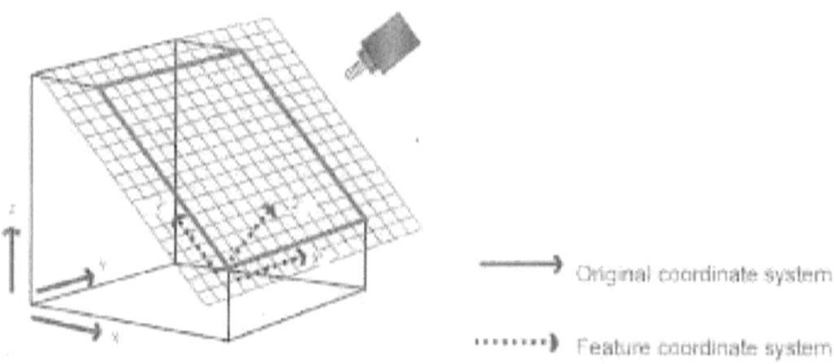

2.32.7.2 Inclined Surface Machining Mode ON (define using Euler angles) (P0 can be omitted)

G68.2 P0 X__ Y__ Z__ I__ J__ K__ ;

Address	Description
X, Y, Z	Feature coordinate system's origin. Command by the absolute values with respect to the coordinate system before issuing the inclined surface machining command.
I, J, K	Euler angles (-360.0° to 360.0°).

Note:

1. If the address **X**, **Y**, or **Z** is omitted, it will be regarded as zero. When all three addresses (**X**, **Y**, and **Z**) are set to "0", the origin of the feature coordinate system will coincide with that of the coordinate system before the inclined surface machining command is issued.
2. If the address **I**, **J**, or **K** is omitted, it will also be regarded as zero.
3. If any address other than **P**, **X**, **Y**, **Z**, **I**, **J**, or **K** is included, a program error (**P954**) will occur.

Detailed Description

When commanding **G68.2 P0** (definition using Euler angles), the feature coordinate system—formed by rotating and shifting the origin of the coordinate system prior to inclined surface machining—is defined.

The rotation of the coordinate system is based on **Euler angles.**

Example:

When the command

G68.2 Xx Yy Zz Ia Jb Kc;

is executed, the feature coordinate system is established as follows:

a) Define a point (**x**, **y**, **z**) in the coordinate system before issuing the inclined surface machining command as the origin of the feature coordinate system.

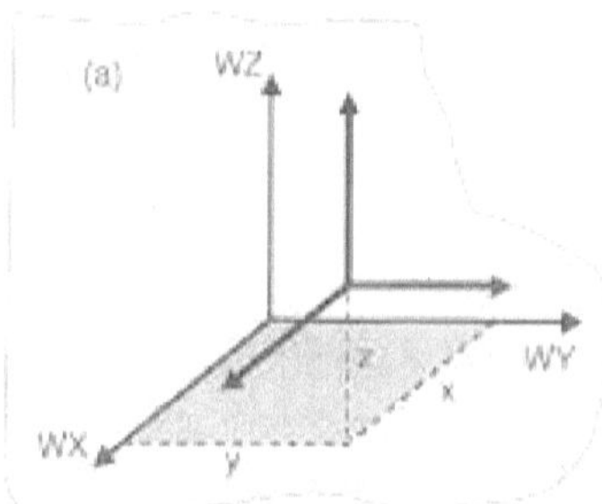

b) Rotate the coordinate system defined in step (a) by angle **a** about its **Z-axis.**

If the coordinate system rotates counterclockwise when viewed from the positive end of the rotation center axis, this rotation is regarded as a forward rotation

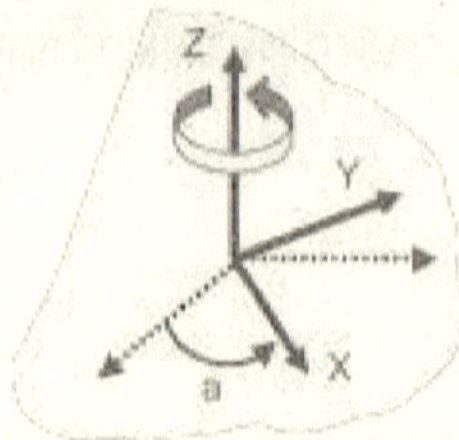

c) Rotate the coordinate system obtained in step (b) by angle **b** about its **X-axis**.

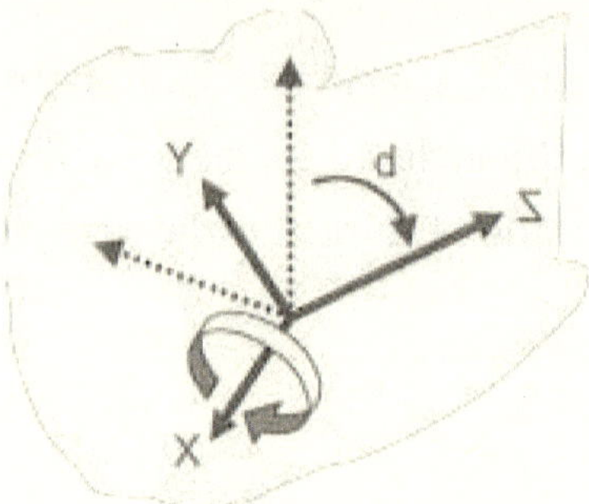

d) Rotate the coordinate system obtained in step (c) by angle **c** about its **Z-axis**.

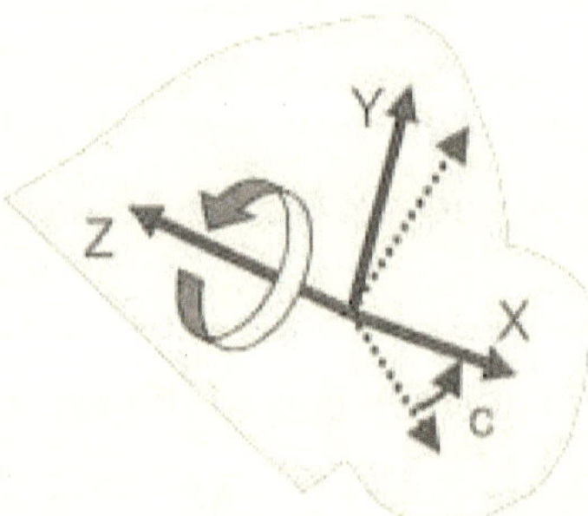

e) The coordinate system resulting from the above operations becomes the **feature coordinate system**.

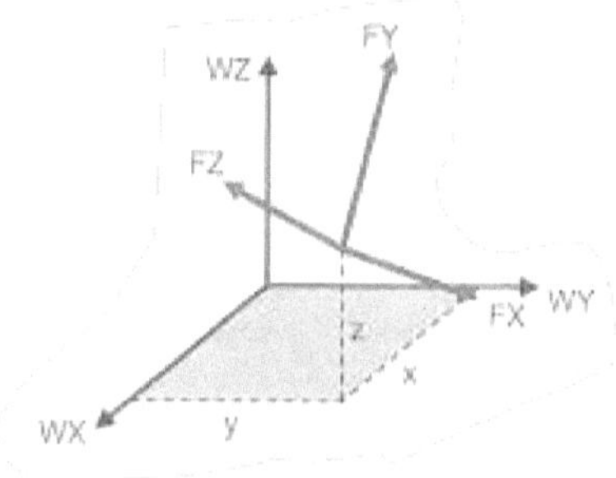

After invoking G68.2, the control rotates the coordinate system accordingly, and all programmed motions are interpreted in the newly tilted plane. To cancel the rotation, the **G69** command is used.

2.32.7.3 Example — Tilted Plane Machining

Example 1:

Machine with A & C configuration

Machine a hole on a plane tilted 90° around the X-axis.

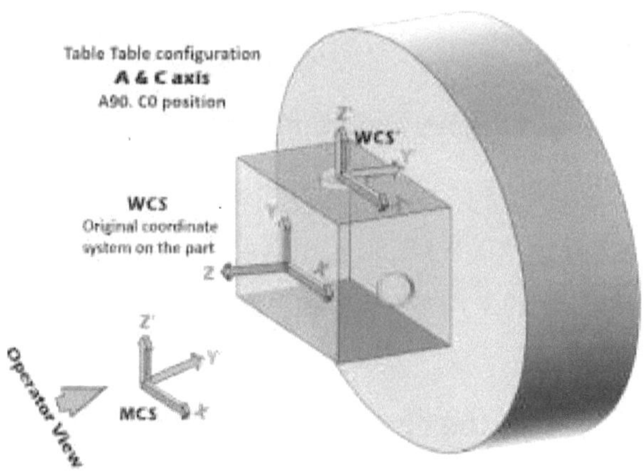

```
G90 G17 G40 G80 G54

G0 A0 C0

G0 X0 Y0 Z100.

G01 A90. C0. F10000.

G68.2 X0. Y30. Z-20. I0. J-90. K0.   (Tilt plane)

(If X, Y & Z are zero, zero point will be

 same as that of original coordinate system)

G53.1                      (Switch to tilted plane mode)

G0 X0 Y0 Z5.

G81  Z-10. R2. F100.

G80

G69                        (Cancel coordinate rotation)
```

Explanation:

G68.2 defines the new coordinate system by rotating 90° about X.

Before tilting, X0, Y0 & Z0 is moved to new point (center of pocket) with X in same position, Y moved by 30 & Z moved by -20

G53.1 activates the **tilted working plane mode**, so subsequent coordinates are interpreted within that rotated system.

Once the machining is complete, G69 returns the control to the original coordinate system.

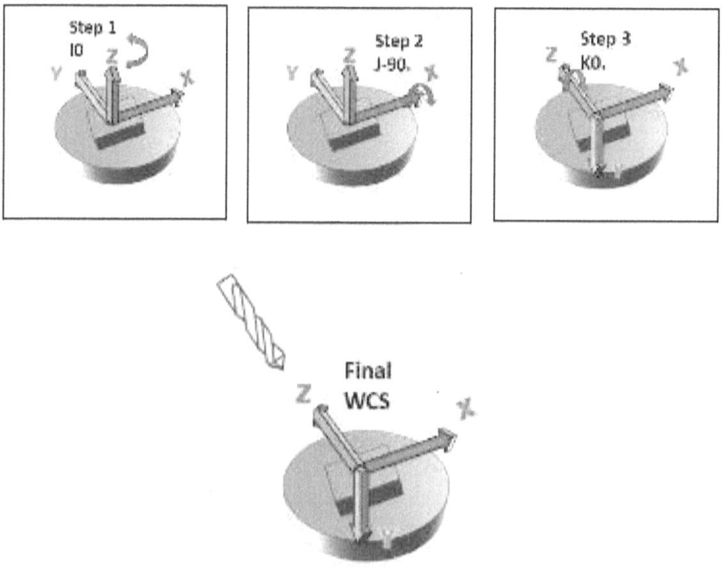

Example 2:

Machine with B & C configuration

hole on a plane tilted 90° around the Y-axis.

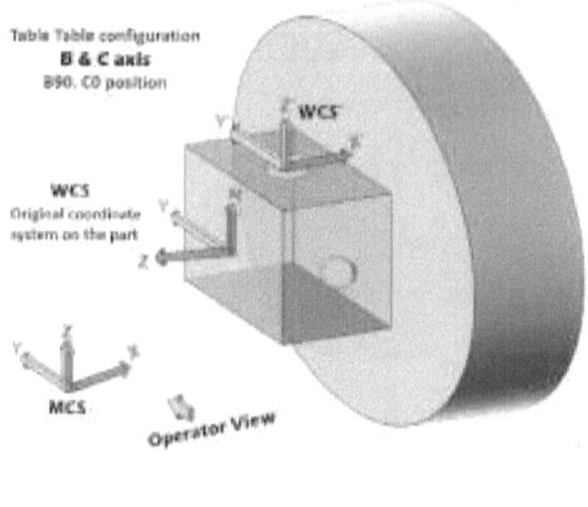

```
G90 G17 G40 G80 G54

G0 B0 C0

G0 X0 Y0 Z100.

G01 B90. C0. F10000.

G68.2 X30. Y0. Z-20. I90. J90. K-90.   (Tilt plane)

G53.1            (Switch to tilted plane mode)

G0 X0 Y0 Z5.

G81 Z-10. R2. F100.

G80

G69              (Cancel coordinate rotation)
```

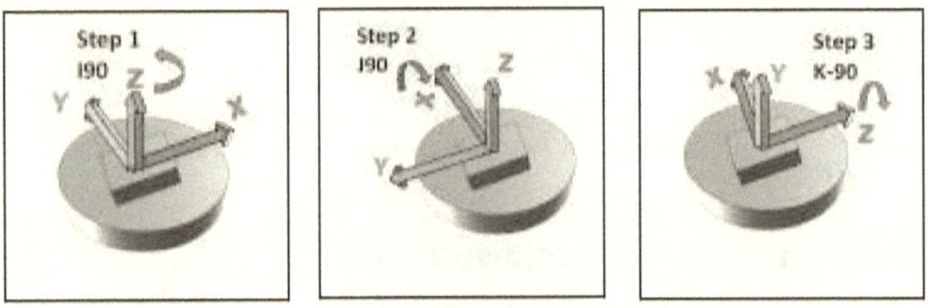

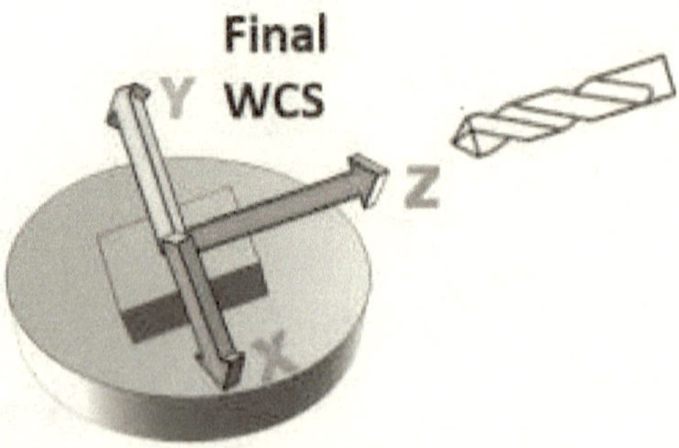

2.32.8 Conceptual Understanding

The G68.2 command mathematically transforms the coordinate system. This means:

The toolpath can be written as if the plane were flat (parallel to XY).

The control internally calculates all required movements to maintain tool orientation.

It greatly simplifies programming for **3+2 machining**, **angled holes**, and **compound surfaces**.

2.32.7.5 Common Applications

Drilling or milling features on **angled faces** without using CAM-generated toolpaths.

Creating **multi-face programs** where the same pattern repeats on tilted planes.

Simplifying **fixture design** by compensating for setup angles through coordinate rotation.

Enhancing **post-processing flexibility** in 5-axis setups.

2.32.7.6 Notes and Precautions

- G68.2 should always be followed by **G53.1** before executing tool motion commands.
- The coordinate rotation is **cumulative**; always cancel it with **G69** before applying a new one.
- Verify the direction of rotation

- Plane selection (G17/G18/G19) is ignored once G68.2 is active, as the system redefines its own tilted plane.
- Available only on controls that support **Tilted Working Plane Function** (e.g., Fanuc 30i/31i/32i series).

2.32.7.7 Key Difference: G68 vs G68.2

Function	Plane Type	Rotation Axes	Use Case
G68	2D (XY, XZ, or YZ)	Single-axis rotation	Simple 2D angular features
G68.2	3D (Tilted plane)	Multiple axes (X, Y, Z)	5-axis positioning and angled machining

2.32.7.8 Key take away

The G68.2 command elevates traditional coordinate rotation to three-dimensional space, giving machinists and programmers a powerful tool for **tilted plane machining**. It eliminates the need for complex trigonometric setups, reduces programming time, and ensures geometric accuracy in multi-axis operations.

When properly combined with **G53.1** and **G69**, G68.2 becomes a cornerstone of **3+2 programming**, bridging the gap between 3-axis simplicity and full 5-axis capability.

2.33 Programmable Offset Setting – G10

2.33.1 Introduction

The **G10 command** in CNC programming is used to **programmatically write, modify, or preset work offsets, tool length offsets, and other parameter values** directly from within a CNC part program.
This eliminates the need for manual data entry through the machine's offset pages, allowing full automation and consistency across setups.

2.33.2 Purpose and Advantages

Using G10 is especially beneficial in production environments where:

- Fixtures or parts are loaded repeatedly on multiple machines.
- Work offsets must be adjusted automatically without operator intervention.
- Tool length or wear data must be standardized across setups.

- Probing or automation systems need to update offsets dynamically.

By incorporating G10, the CNC program can define or update the same offsets each time it runs—ensuring precision and eliminating human error.

> CAUTION
>
> If this function is not carefully used, offset data can get relaced and results can be disastrous

2.33.3 For work offsets

G10 L2 Pp IP_;

p=0 : External workpiece origin offset value
p=1 to 6 : Workpiece origin offset value corresponding to workpiece coordinate system 1 to 6
IP_ : For an absolute command, workpiece origin offset for each axis.
For an incremental command, value to be added to the set workpiece origin offset for each axis (the result of addition becomes the new workpiece origin offset).

Examples

1. Setting Work Offset (G57)

G10 L2 P4 X100.000 Y200.000 Z-50.000

- Writes X = 100.000, Y = 200.000, and Z = –50.000 into the **G57** offset register.
- Equivalent to manually typing those values on the "Work Offset" screen.

2. External Work Offset

G10 L2 P0 X0.0 Y0.0 Z50.0

- Writes Z = 50.000 into the external work offset register.
- Equivalent to manually typing those values on the "Work Offset" screen.

2.33.4 Specific for Machining center

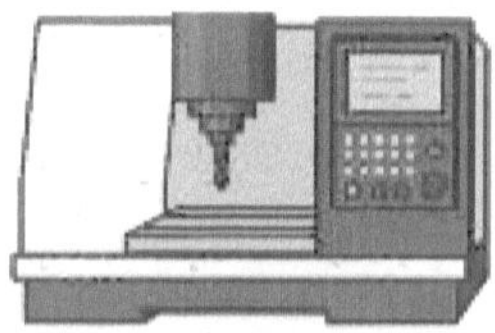

For tool compensation memory A

G10 L11 P_ R_ Q_ ;

P_ : Tool compensation number
R_ : Tool compensation value
Q_ : Imaginary tool nose number

For tool compensation memory C

G10 L_ P_ R_ Q_ ;

L_ : Type of compensation memory

L10 : Geometry compensation value corresponding to an H code

L11 : Wear compensation value corresponding to an H code

L12 : Geometry compensation value corresponding to a D code

L13 : Wear compensation corresponding to a D code

L110 : Geometry compensation value corresponding to a D code (for corner R offset)

L111 : Wear compensation corresponding to a D code (for corner R offset)

P_ : Tool compensation number
R_ : Tool compensation value
Q_ : Imaginary tool nose number

Example:

Setting tool length offset on tool#5

G10 L11 P5 R50.0 Q1

2.33.5 Specific for Turning center

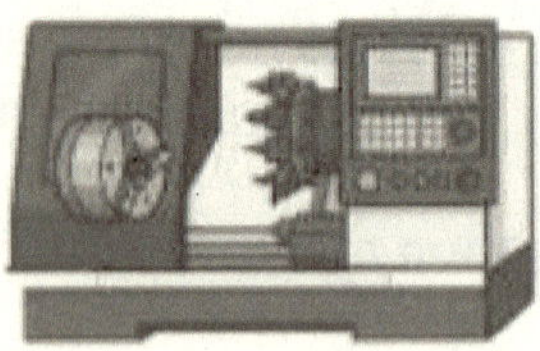

Format:

G10 P_ X_ Y_ Z_ R_ Q_ E_ F_ ;
or
G10 P_ U_ V_ W_ C_ Q_ E_ F_ ;

P : Offset number

0 : Specifies a workpiece coordinate system shift value.

1 to 999 : Specifies a tool wear offset value.

10000+(1 to 999) : Specifies a tool geometry offset value, with a number (1 to 999) representing an offset number.

X : X axis offset value (absolute)
Y : Y axis offset value (absolute)
Z : Z axis offset value (absolute)
U : X axis offset value (incremental)
V : Y axis offset value (incremental)
W : Z axis offset value (incremental)
R : Tool nose radius compensation value (absolute)
C : Tool nose radius compensation value (incremental)
Q : Virtual tool nose number
E : 4th axis offset value (absolute)
F : 5th axis offset value (absolute)

Example:

Writing offset data to Tool#05

G10 P05 X_ Z_ (for wear data for tool#5)

G10 P10005 X_ Z_ (for geometry data for tool#5)

Advanced Use with Macros

G10 is often combined with **macro variables** for flexible automation.

For more details on macro, please refer chapter10.
Example of a probing cycle:

#100 = #5063 (Capture measured X position)

#101 = #5069 (Capture measured Z position)

G10 L2 P1 X#100 Z#101

Here, the machine measures a feature and automatically writes the measured values to the **G54 offset**, ensuring precise, automatic alignment after each setup.

Precautions

- Ensure the correct **L** and **P** values are used—incorrect entries may overwrite critical offsets.
- Always verify the coordinate system currently active before executing a G10 command.
- G10 changes remain **permanent** until rewritten or cleared.

PRECISION RECAP

Mastering CNC programming requires a deep understanding of both the theoretical foundations and the practical execution of G-codes and M-codes. Core concepts such as coordinate systems, linear, circular, and helical interpolation form the foundation for precise and efficient toolpath generation. Effective use of Cutter Radius Compensation (CRC) enables machinists to program directly from part geometry while maintaining dimensional accuracy and flexibility. Understanding absolute and incremental dimensioning is fundamental to controlling motion reference and preventing cumulative errors during complex machining sequences. Proper plane selection ensures correct circular interpolation, CRC direction, and infeed logic for cycles. Canned cycles and subprograms improve modularity, reduce repetition, and enhance overall program reliability. Basic turn-mill centers synchronize C-axis rotation with X-axis motion—though the absence of a Y-axis limits versatility—making simulation, verification, and process safety indispensable for achieving consistent, high-precision results. The coordinate rotation command empowers programmers to perform dynamic work plane transformations in 3D space—an essential feature for 5-axis and multi-face machining where tool alignment and orientation accuracy are critical.

Chapter 3: Tooling, Work holding & Setup

CNC FOCUS POINTS

By the end of this chapter, you will be able to:

- **Select optimal cutting tools and holders** tailored to specific materials, geometries, and machine capabilities.
- **Design robust work holding solutions** that maximize rigidity, minimize vibration, and ensure repeatability during both prototyping and production.
- **Execute precise, error-free setups** to avoid costly mistakes and wasted material.
- **Troubleshoot common issues** using practical templates, diagnostic checklists, and a deeper understanding of the machining process.

3.1 Tooling Fundamentals: Foundations of Success

Golden Rule

Always match tooling to the planned operation, material, and machine characteristics. Only once compatibility is ensured should you optimize for tool life, productivity, and finishing performance.

3.1.1 Quick Tool Selection Table

Operation	**Recommended Tool**	**Key Notes**
Facing	45° Face Mill	Positive rake for aluminum/stainless
Slotting (T-slots)	3–4 Flute End Mills	Key seat cutters / T-slot cutters
Profiling/HEM	5–7 Flutes (steel), 3–4 (aluminum)	Lower radial DOC in HEM
Rough Pockets	Variable-Helix, Variable-Pitch Rougher, Corncob (serrated rougher)	Maximize MRR

Operation	Recommended Tool	Key Notes
Finishing	4–6 Flutes, Unequal Pitch	Stepover 5–15% of tool diameter
Drilling	Carbide (<12mm), Indexable (>12)	Parabolic flutes; peck for chip control
Tapping	Spiral Point/Flute, Form Tap	Match tap type to hole & material

For more details on cutting tools, refer chapter 5

3.1.2 Material-Specific Guidelines

- **Aluminum**: Sharp geometries, high rake, polished flutes, broad chip evacuation.
- **Steel**: Edge hone for toughness, 4–6 flutes.
- **Stainless/Titanium**: Strong cutting edges, slow SFM, use heavy coolant, heat-resistant coatings.

3.1.3 Critical Tool Setup Parameters

- **Stick-Out Ratio**: ≤4×D for best stability; >6×D needs reduced parameters or special holders.

- **Runout (TIR)**: Roughing ≤10–15 µm; finishing and micro-tools ≤3–5 µm for superior finish.
- **Inspection**: Always inspect for edge chipping, wear, or buildup before use.

3.2 Tool Holding Systems: Precision Meets Practicality

Tool holders transmit cutting forces, influence vibration, runout, and must match tool and operation requirements.

3.2.1 Holder Types

Holder Type	**Runout (µm)**	**Strengths**	**Applications**
REGO-FIX powRgrip	1–3	Ultra-low runout, high force	Long-reach, finishing
Premium ER Collets	8–15	Versatile, economic	General-purpose
Hydraulic Chucks	3–5	Excellent damping, low runout	Finishing, vibration-prone

Holder Type	Runout (µm)	Strengths	Applications
Shrink-Fit	2–4	High speed, slim profile	High-RPM, tight spaces
Power/Milling Chucks	10–20	High torque	Heavy roughing, large tools
Side-Lock (Weldon Flat)	15–25	Pull-out resistance	Aggressive roughing only

Anti-Pull-Out Tip: Use helical-groove shanks for high-axial-force HEM.

3.2.2 Best Practices

- Balance assemblies for >12,000 RPM.
- Always clean tapers/holders before use.
- Measure runout at the tip using a dial indicator.
- Inspect monthly for fatigue, cracks, or wear

3.3 Work holding Mastery: The 3-2-1 Principle

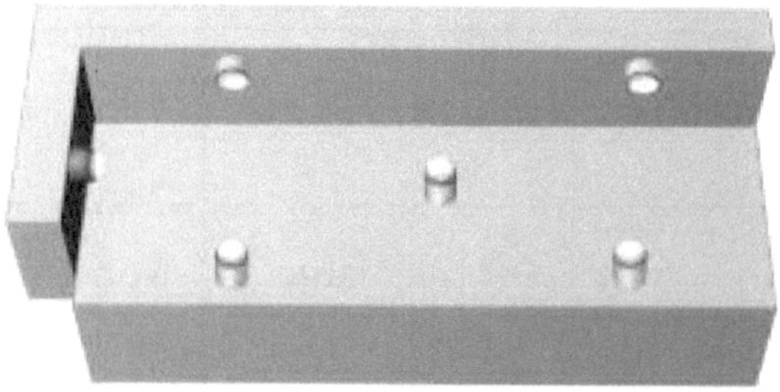

3.3.1 Constrain All Six Degrees of Freedom

- **3 points**: Establish primary plane (typically base).
- **2 points**: Fix secondary plane (usually part side).
- **1 point**: Secure remaining axis with an end stop.

Refer chapter 6 for more details

Method	Strengths	Applications
Vise + Soft Jaws	Customizable, repeatable	General small/medium parts
Strap/Toe Clamps	Low profile, strong	Odd shapes, heavy blocks

Method	Strengths	Applications
Modular Plates/Tombs	Multi-part setups	Production, 4th-axis
Vacuum Fixtures	For thin, large plates	Non-ferrous, aerospace
4th-Axis Fixtures	Multi-face access	Rotational parts, complex

Tips

- Machine soft jaws in-situ for zero lift.
- Always remove chips, debris from surfaces before setup.

3.4 Setup Planning: Think Before You Act

3.4.1 Setup Sheet Must-Haves

- Part info: number, revision, quantity, material, operation index.
- Datum strategy: Clearly define work offset (G54+), datum location.
- Tool list: Tool, holder, stick-out, unique notes.

- Work holding plan: Sketches/photos, clamp sequence, torque settings.
- QC strategy: List critical features, inspection methods, checkpoint frequencies.
- Program management: Simulation status, file names, revision history.

3.4.2 Coordinate System Philosophy

- Top of stock: Simpler for one-time parts.
- Finished surface: Best for multi-op parts, tight tolerances.
- Fixture pad: Ideal for production with modular systems.

Tip: Always document everything—photos, notes, and correction sheets speed up troubleshooting and onboarding.

3.4.3 Machine Preparation: Foundation of Precision

- Vise alignment: ≤0.01–0.02 mm parallel using a dial indicator.
- Spindle tramming: <0.01 mm over 200 mm circle for finishing.
- Stock prep: Flat, parallel faces; leave machining allowance for cleanup.

- Machine warm-up: 10–15 minutes of spindle/axis cycling.

3.4.4 Work Offsets & Tool Length: Digital Accuracy

- Set work offsets (G54–G59) with touch probes/edge finders; verify with test indicators.
- Tool length offsets: Reference every tool to the same setup plane or pad.
- Apply and verify G43 H-codes for tool length compensation.
- **Validation workflow:** CAM sim → Dry run at Z+50 mm → Single-block first tool → Confirm clearance on approach/retract.

For more details on work offsets and tool offsets, refer chapter 4

3.4.5 Coolant & Chip Management

- Flood coolant: Universal; manages heat, chip evacuation.
- Through-spindle coolant: For high aspect ratio drilling, tough alloys.
- Re-align nozzles post-tool-change to ensure proper coolant reach.
- Confirm coolant mix (typically 5–10%) to avoid rust, bacteria.

- Cast iron: Prefer dry machining or mist to avoid sludge.
- Use active chip conveyors in production; hand-clear chips if needed.

3.4.6 Feeds & Speeds: Calculating the Baseline

- **Spindle Speed:** $N = \frac{1000 \times V_c}{\pi D}$
- **Feed Rate:** F=fz × z × n
    - *Vc* = cutting speed (m/min), *D* = tool diameter (mm), *fz* = feed per tooth (mm/tooth), *z* = flutes, *n* = RPM.

Material	Feed per Tooth (mm)
Aluminum	0.03–0.10
Steel	0.02–0.06
Stainless	0.01–0.04

- For HEM, use radial DOC = 5–15% of tool diameter, axial DOC = 1–2×D.

Tip: Start conservatively—observe spindle load, chip form, and surface quality, then adjust.

For more details on machining strategies, refer chapter8

3.5 Shop Standards

- Tool runout: Roughing ≤15 µm; finishing ≤5 µm.
- Vise jaw alignment: ≤0.02 mm over 100 mm.
- Rotary indexer repeatability: 0.01–0.02 mm.
- Clamping torque always per manufacturer; check after heavy cuts.

3.6 Troubleshooting Guide

Issue	**Likely Cause**	**Solution**
Chatter	Tool stick-out, weak holding	Reduce stick-out, use damping holders, adjust RPM/feed
Poor finish	High runout, dull tool, wrong speeds	Check runout, sharpen tool, tweak SFM & coolant
Tool pull-out	Poor grip, aggressive DOC	Use anti-pull-out, reduce DOC/WOC, check torque

Issue	Likely Cause	Solution
Part movement	Weak fixturing, inadequate bite	Increase jaw engagement, re-clamp, clean all fixtures
Burrs	Low chip load, worn tool	Boost chip load, use sharp tools, spring pass, deburr

3.7 Ready-to-Use Templates

3.7.1 Tool List Table Example

Tool No	Description	Holder	Stick-Out	Notes
T01	Ø20 Face Mill	Hydraulic	45 mm	Finish pass only
T02	Ø10 Carbide End Mill	Shrink-Fit	28 mm	Rough/ HEM, 10% WOC

Tool No	Description	Holder	Stick-Out	Notes
T03	Ø6 Carbide End Mill	ER16	20 mm	Finish walls
T04	Ø6 Drill	ER16	35 mm	Through-coolant
T05	M6 Spiral Tap	Tension/Comp	—	Blind holes, low RPM

3.7.2 Setup Sheet Elements

- Part details: number, revision, material, quantity.
- Coordinate system: Datum type, e.g., G54 on part surface.
- Work holding: Layout sketch, order of clamping, torque values.
- Tool list: Stick-out, holder type, operation notes.
- QC Plan: Highlight tight-tolerance features and inspection methods.

- First Article Approval: Sign-off, date, QC checks.

3.8 Hands-On Exercises

Runout Audit: Measure runout for collet, hydraulic, and shrink-fit holders; compare part finishes.

Soft Jaw Design: Machine and test soft jaws, perform "tug test," compare to smooth jaws for securement.

Offset Repeatability: Set G54 via probe and edge finder; record values for repeatability analysis.

PRECISION RECAP

This chapter provides the essential knowledge to achieve reliable, accurate CNC machining setups. By mastering tool and holder selection, work holding design, and methodical setup planning, you'll eliminate common errors, increase efficiency, and consistently deliver high-quality components. Use the included templates and troubleshooting tables as ongoing shop references—building confidence and capability with every part, you machine.

Chapter 4 — Mastering Offsets

CNC FOCUS POINTS

In this chapter, you will learn how CNC machines understand where a part is located and how tools reach the correct cutting position. You'll discover the difference between the machine's own zero point and the part zero you set for machining. You will understand how to use Work Coordinate Systems (like G54), set tool heights using tool length offsets (G43), adjust cutter radius for accuracy (G41/G42), and use probing systems for automatic measurement. By the end, you will be able to set up a job confidently, choose the right offset method, and make precise adjustments without changing the program — a key skill for every professional CNC machinist

4.1 Understanding CNC Offsets

4.1.1 Introduction:

Every CNC machine operates within two coordinate worlds — its own mechanical reality and the part's coordinate system.

The machine has a permanent reference, called **Machine Zero**, fixed by the manufacturer. In contrast, your part program references **Part Zero**, which may lie anywhere on the table, vise, or fixture.

So, the fundamental question arises: *How does the machine know where the part actually is?*
The answer lies in **offsets** — the translator between these two coordinate worlds. Offsets instruct the machine:

"For this setup, start measuring from here — not from your home position — and adjust for each tool's reach accordingly."

Offsets enable a single program to run accurately regardless of workpiece placement, making CNC machining repeatable, flexible, and efficient.

4.1.2 The Machine Coordinate System (MCS)

The **Machine Coordinate System (MCS)** acts as the machine's internal reference grid. Its origin, the **Machine Zero Point**, defines the limits of motion for axes X, Y, and Z.
All homing sequences and calibrations are referenced to this zero point.

While critical for operation, MCS isn't practical for part programming because workpieces rarely align with Machine Zero.

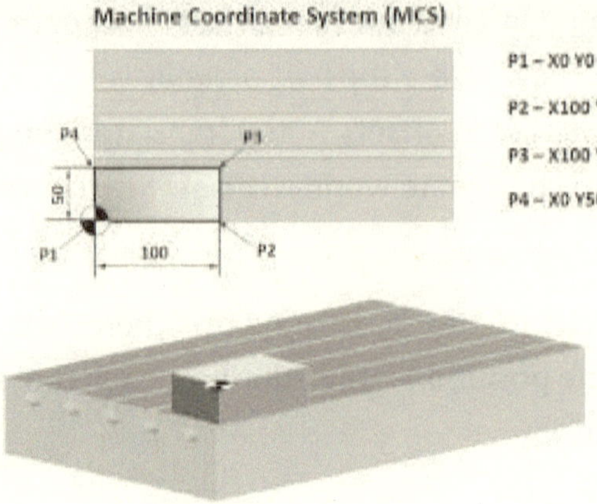

4.1.3 Programming with MCS

Imagine programming a part using the lower-left corner as X0 Y0. On the machine, that point might actually lie at X = 90.123 mm and Y = 60.45 mm from Machine Zero.

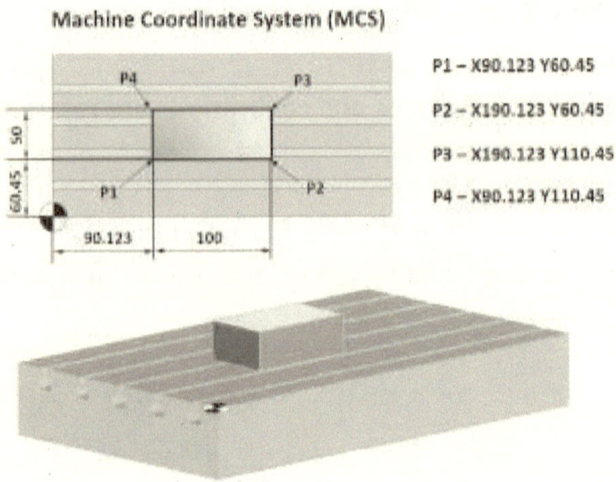

If you relied on MCS alone, every coordinate in your program would need those offsets manually added — tedious and error-prone.

Without Offsets (Manual Calculation)	With Work Coordinate System (Simplified)
P1 – X90.123 Y60.45	P1 – X0 Y0
P2 – X190.123 Y60.45	P2 – X100 Y0
P3 – X190.123 Y110.45	P3 – X100 Y50
P4 – X90.123 Y110.45	P4 – X0 Y50

With a **Work Coordinate System** (e.g., G54) active, the controller automatically adds these stored distances.

Your G-code stays identical to the drawing, while the machine handles all translations internally — the foundation of modern CNC accuracy.

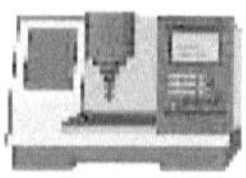

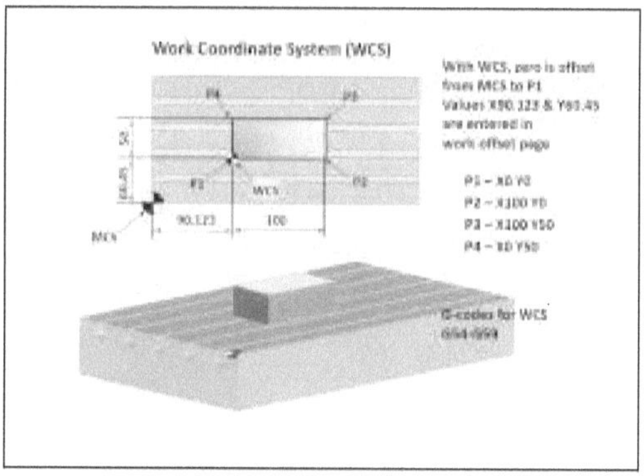

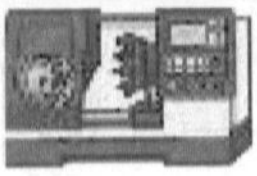

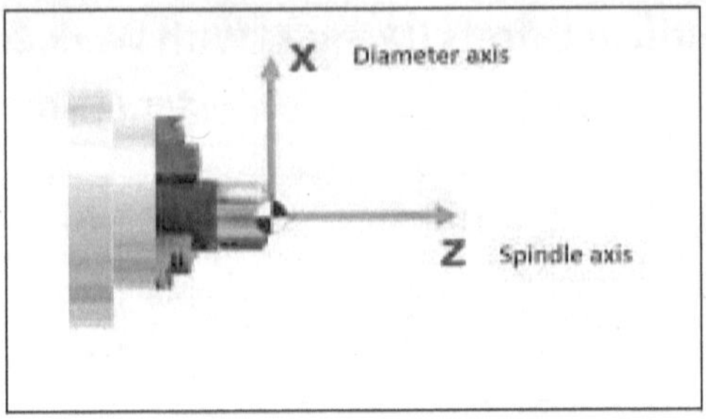

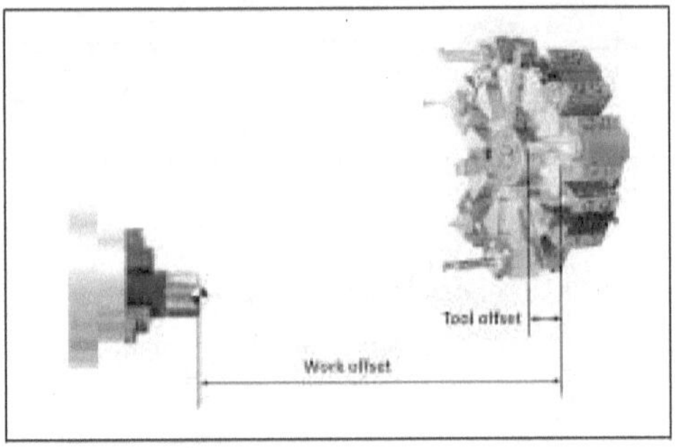

4.1.4 Tool Length Variation

After X and Y are set, the Z-axis introduces another challenge: every tool (long drill, short end mill, chamfer mill) has a unique length.
Without correction, a single command like Z0 positions tool tips at different heights — risking gouges or uncut material.

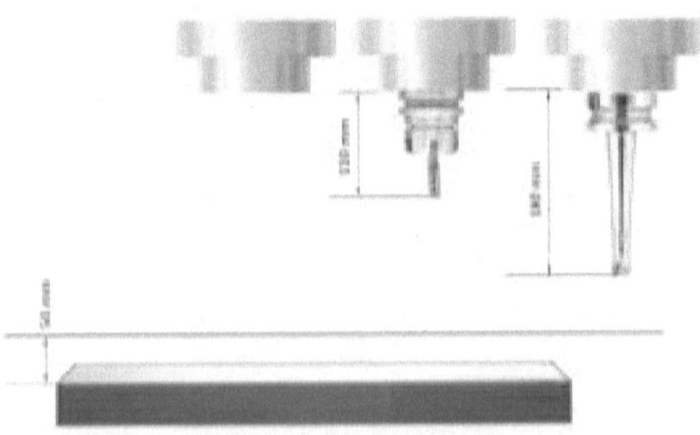

Tool Length Compensation (TLC) stores each tool's length in the controller.
When invoked (e.g., G43 H1), the control compensates automatically so every tool references the same Z-zero plane.

4.2 Work Coordinate System (WCS)

4.2.1 Key Concepts

- **Machine Zero:** Factory-set reference at the home position
- **Part Zero:** User-defined origin on the workpiece or CAD datum
- **Work Offset Values:** Stored distances between Machine Zero and Part Zero

Purpose:
WCS allows parts to be positioned freely within the machine's travel while keeping programs unchanged. Without WCS, fixtures would require identical placement each time — a major inefficiency.

4.2.2 Standard Work Coordinate Systems

- **G54 – G59:** Primary coordinate systems for most setups
- **G54.1 P1–P48 :** Extended sets for multi-fixture or multi-part jobs

Example:

```
G54             ; Select WCS 1
G00 X0 Y0 Z50   ; Rapid to part zero at safe height
G01 Z-5.0 F100  ; Feed to cutting depth
```

4.2.3 Applications

- Multi-vise or multi-fixture setups (e.g., G54 = Vise 1, G55 = Vise 2)
- Tombstone machining using multiple WCS
- Batch production reusing identical programs across setups

4.2.4 Setting Work Offsets

Method	**Accuracy**	**Notes**
Edge Finder	± 0.01 mm	Manual, economical
Dial Indicator	± 0.005 mm	Ensures parallelism and alignment
Tool Touch-Off	± 0.01 mm	Common for Z-zero
Touch Probe	± 0.001 mm	Fast, automated (Renishaw MP250)
Laser / Vision	Non-contact	Ideal for delicate parts

4.3 Tool Length Compensation (TLC)

4.3.1 Importance

Tool lengths vary due to holder type, geometry, or wear. TLC ensures programmed Z-coordinates remain consistent across all tools by compensating for each tool's measured length.

4.3.2 Key Codes

G-Code	Function
G43 H__	Apply tool-length compensation
G43.1	Dynamic real-time adjustment (control-specific)
G49	Cancel tool-length compensation

Example:

```
T01 M06         ; Load tool #1
G43 H01 Z50     ; Apply offset #1, move to safe height
M03 S1200       ; Spindle ON
G01 Z-10 F150   ; Feed to depth
```

4.3.3 Measuring Tool Length

Method	Accuracy	Notes
Offline Presetting	± 0.002 mm	Fast, consistent in toolroom control
On-Machine Probing	± 0.002 mm	Verifies in situ during setup
Manual Touch-Off	Variable	Simple but operator-dependent

Best Practice: Preset tools offline for speed, verify critical ones on-machine, and monitor wear regularly.

4.3.4 Offline tool presetting

A tool presetter is a precision device used to measure and set tool dimensions—such as length and diameter—offline before tools are loaded into a CNC machine. By pre-measuring tools, it reduces spindle downtime, improves machining accuracy, and ensures consistent setup quality. Tool presetters enhance productivity, minimize human error, and support efficient, repeatable CNC operations in modern manufacturing.

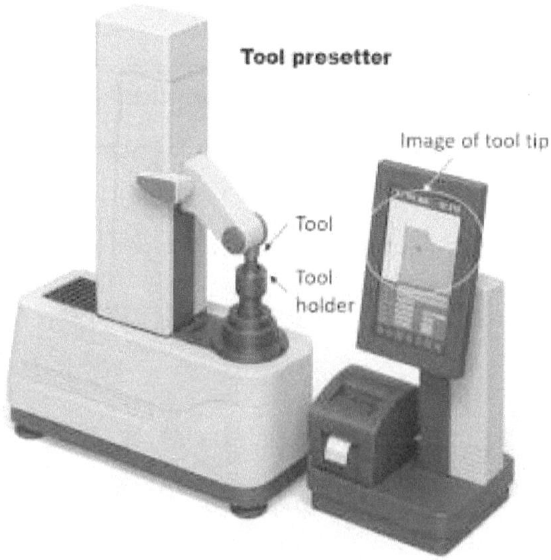

4.4 Cutter Radius Compensation (CRC)

4.4.1 Benefits

- Adjusts toolpath to match cutter radius without reprogramming
- Compensates automatically for wear or regrind
- Enables accurate contouring and stock allowance adjustment

4.4.2 Key Codes

G-Code	Function
G40	Cancel CRC
G41	Compensate Left (Climb Milling)

G-Code	Function
G42	Compensate Right (Conventional Milling)

Refer chapter2 section 2.16 for more details on CRC

4.5 Automated Offset Setting (Probing)

4.5.1 Workpiece Probing

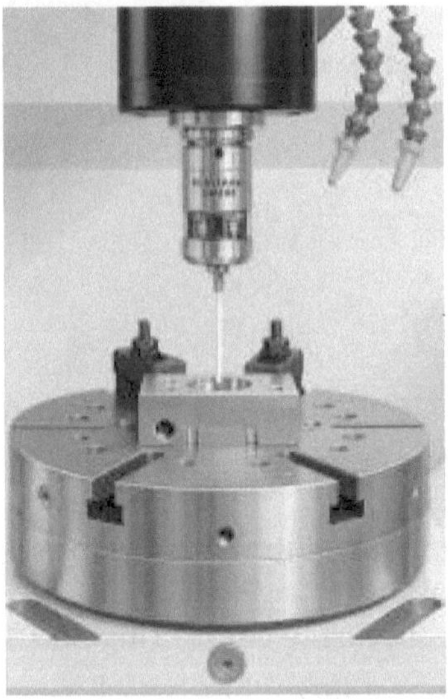

High-precision probes (Eg: Renishaw, Blum) achieve ± 0.001 mm accuracy.
They automate edge finding, Z-height measurement, hole-center detection, and part alignment.
Probes can directly update work offsets using commands like:

```
G10 L2 P1 X0 Y0 Z0
```

(Refer Chapter2 section 2.33 for more details on G10)

4.5.2 Tool Probing

Laser, contact, or optical systems measure tool length, diameter, wear, and detect broken tools (± 0.002 mm).

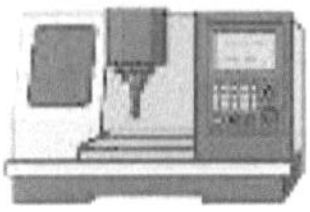

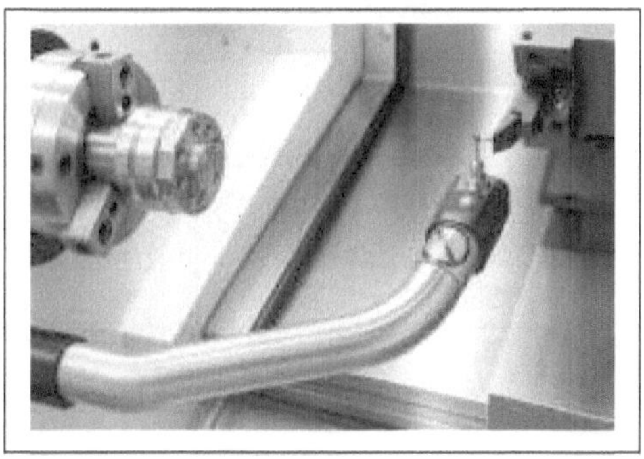

4.5.3 Integration

When probe data is integrated with MES/ERP systems, offset values become dynamic feedback for process control.
Track **SPC metrics**, tool life, and offset drift for predictive maintenance.

4.6 Setup Workflow Example

A robust offset workflow follows these ordered steps:

1. Review drawing and tolerances
2. Preset tools; confirm fixture stability
3. Load program and set G54 (X/Y/Z) using probe or edge finder
4. Record offsets in controller table
5. Apply G43 H__ for each tool
6. Enable G41/G42 for contouring
7. Run first part in single-block mode
8. Inspect dimensions; fine-tune offsets if required

4.7 Troubleshooting Offsets

Issue	**Cause**	**Fix**
Wrong H/D Register	Offset mismatch	Standardize numbering; label tools clearly

Issue	Cause	Fix
Forgotten Cancel	G49/G40 missing	Always cancel before tool change or program end
Probe Errors	Incorrect approach path	Use standardized probing cycles
Tool Length Error	Inconsistent touch-off	Calibrate presetters; follow documented method

Checklist

- Verify offset registers (H, D, G54–G59)
- Cross-check tool library data
- Calibrate probes regularly
- Monitor offset drift via SPC for thermal or wear trends

4.8 Advanced Offset Applications

4.8.1 Multi-Part Machining

- **Tombstone Machining:** Assign different WCS codes (G54–G57) for multi-part production.
- **Part Families:** Use variable offsets within one program for similar parts.

4.8.2 Coordinate Transformation

- **G68/G69:** Rotate coordinate systems for angled features (Refer chapter2 section 2.32.6 for details)
- **Adaptive Machining:** Dynamically update offsets from probe feedback

PRECISION RECAP

Offsets are the unseen backbone of CNC precision. They connect digital instructions to real-world movement — enabling fast setups, reliable accuracy, and flexible production.

Key Takeaways

- Use **WCS (G54–G59)** to establish part origins
- Apply **TLC (G43/G49)** to account for tool-length variations
- Use **CRC (G41/G42)** to maintain dimensional accuracy across tool wear
- Employ **probing** for automated, error-free offset setting

Chapter 5: Cutting Tools Demystified — Geometry, Grades & Performance

CNC FOCUS POINTS

By the end of this chapter, you will:

- Understand how end mill geometry—including flute count, helix angle, rake angle, relief angle, and edge preparation—affects cutting forces, surface finish, and tool life.
- Select suitable carbide grades and coatings for materials such as steel, stainless steel, aluminum, titanium, cast iron, and superalloys.
- Choose the correct tool profile (e.g., square end, corner-radius, ball, chip breaker, high-feed) for specific machining operations.
- Apply practical starting parameters for surface feet per minute (SFM), chip load, and radial and axial engagement.
- Diagnose and mitigate common tool wear patterns.
- Implement cost-efficient tool strategies balancing performance and economics.

5.1 Tool Anatomy & Geometry Fundamentals

5.1.1 Flute Count Selection

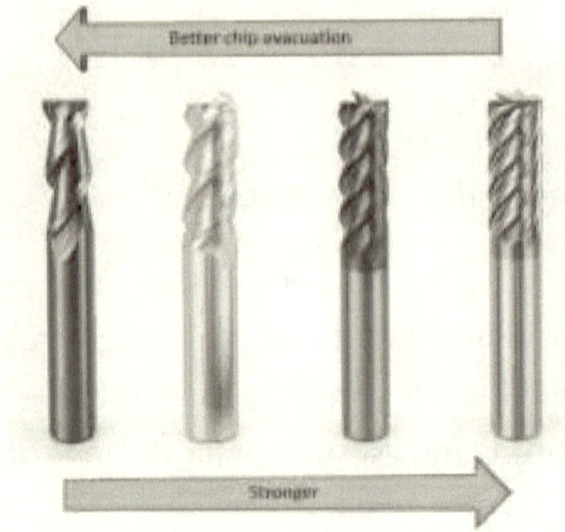

- **Aluminum & Non-Ferrous (2–4 flutes):** Provide large flute valleys to aid efficient chip evacuation, reduce cutting forces, and perform well with thin walls.
- **Steel (4–6 flutes):** Balance rigidity and chip evacuation to achieve smooth finishes and handle interrupted cuts effectively.
- **Stainless/Titanium (5–7 flutes):** Maintain constant cutting edge engagement, reduce chip thickness, and minimize chatter for tough materials.

5.1.2 Helix Angle Optimization

Helix angle

- **Standard Helix (30°–35°):** Works for most materials by balancing cutting forces and chip evacuation.
- **High Helix (40°–45°):** Improves chip removal and surface finish; best for aluminum and stainless steel but requires rigid work holding to avoid vibration.
- **Variable Helix:** Disrupts harmonic resonance, reduces chatter, and balances cutting forces; ideal for high-speed machining.

5.1.3 Rake & Relief Angles

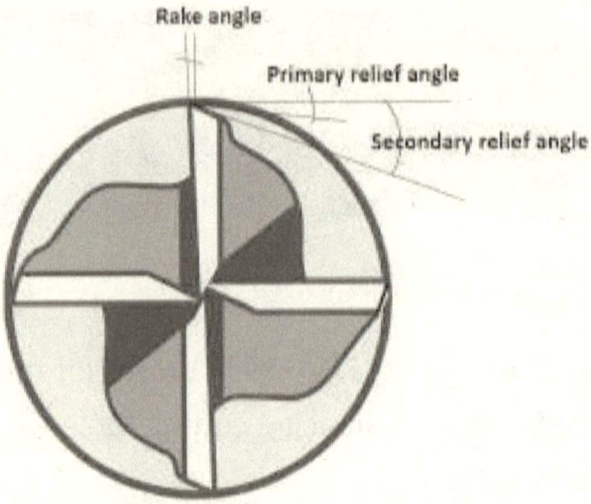

- **Positive Rake:** Creates a sharp cutting edge with low cutting forces, suited for aluminum, plastics, and thin sections. It may chip easily.
- **Neutral/Robust Rake:** Offers stronger edges for steel and stainless steel; increases cutting forces but extends tool durability.
- **Relief Angle:** Balance between too little (causes rubbing and heat) and too much (weakens edges). Aim for 5°–10° for optimal strength and sharpness.

5.1.4 Edge Preparation

- **Sharp Edge:** Minimizes forces, ideal for aluminum and plastics but fragile.

- **Micro-Hone (5–20 µm):** Common for steel; reduces chipping and extends life, slightly increasing cutting forces.
- **Small Chamfer (0.02–0.05 mm):** Strengthens edges for hardened or abrasive materials (>45 HRC), ensuring consistent wear.

5.1.5 End Mill Profiles

- **Square End:** Provides sharp corners essential for slots and pockets; susceptible to chipping; monitor edges closely.
- **Corner Radius (Bull Nose):** Strengthens cutting edges, enhancing durability; leaves a small fillet; versatile for many operations.

- **Ball Nose:** Perfect for 3D contouring and surfacing; requires higher RPMs and small stepovers (3%–8%) due to zero speed at the tip.

5.1.6 Chip breaker Technology

- **Purpose:** Fragment chips for deep slots and pockets to prevent clogging.
- **Benefits:** Lowers cutting forces, prevents chip packing, and supports unattended machining.
- **Applications:** Essential when machining sticky materials such as aluminum or stainless steel and in high-volume shops.

5.2 Application-Specific Tool Selection Guide

Operation	**Recommended Tool**	**Critical Notes**
Face Milling	45° face mill, high-shear inserts	Reduces radial forces; excellent for flat surfaces.
Slotting	3–4 flute with chip breaker	Use trochoidal paths; avoid full-width cuts.

Operation	Recommended Tool	Critical Notes
Profiling/HEM	5–7 flute, variable helix	Small radial (5–15%), deep axial (1–3× diameter).
Wall Finish	4–6 flute, corner-radius, variable pitch	Leave 0.1–0.3 mm stock for finishing passes.
3D Finish	Ball or tapered ball end mill	Use small stepovers (3–8%); avoid tip rubbing.
Thin Walls	Sharp, high-helix, low radial engagement	Climb milling and spring passes improve precision.

5.3 ISO Inserts

5.3.1 Introduction

ISO tool inserts are standardized, replaceable cutting tips used in turning, milling, and drilling operations. Defined under ISO 1832, they follow a uniform coding system that specifies shape, clearance angle, tolerance, chip breaker style, and material grade. This global standardization ensures interchangeability, predictable performance, and easy selection across tool manufacturers.

Each insert is identified by a multi-character ISO code—for example, **CNMG 120408**. In this code:

- **C** indicates an **80° rhombic shape**
- **N** denotes **0° relief (negative rake insert)**
- **M** specifies **tolerance class**
- **G** indicates **chip-breaker and clamping style**
- **12 04 08** refer to insert **size, thickness, and nose radius**

Common materials include **carbide, cermet, ceramics, CBN**, and **PCD**, with coatings like **TiN**, **TiCN**, and **AlTiN** enhancing wear and heat resistance. ISO inserts support consistent machining, reduce downtime via quick indexing, and allow machinists to tailor cutting performance based on material and application.

In modern precision machining, mastering ISO insert codes empowers machinists to choose the right geometry, clearance, and grade—leading to superior surface finish, tool life, and productivity.

5.3.2 Insert Nomenclature

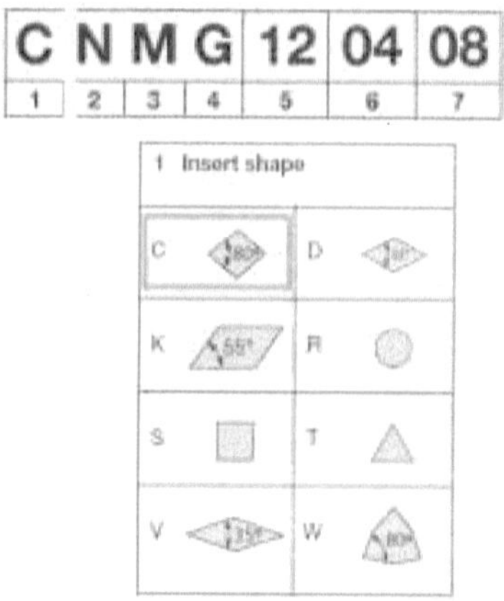

C	N	M	G	12	04	08
1	2	3	4	5	6	7

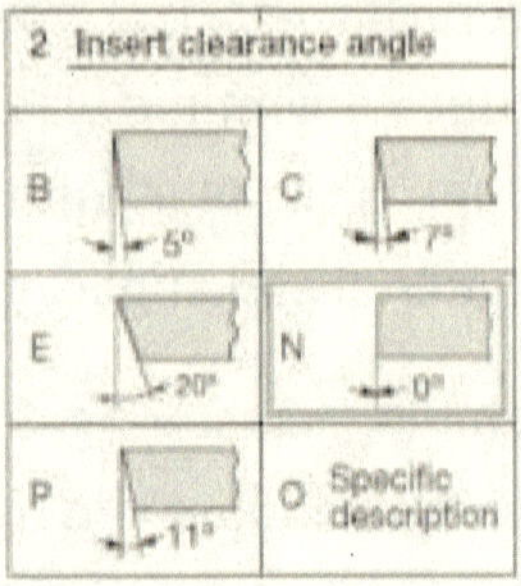

C	N	M	G	12	04	08
1	2	3	4	5	6	7

3 Tolerances

Class	S	IC / W1
G	±0.13	±0.025
M	±0.13	±0.05 – ±0.15[1]
U	±0.13	±0.08 – ±0.25[1]
E	±0.025	±0.025

[1]Varies depending on the size of IC. See below.

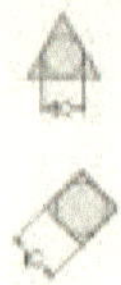

C	N	M	G	12	04	08
1	2	3	4	5	6	7

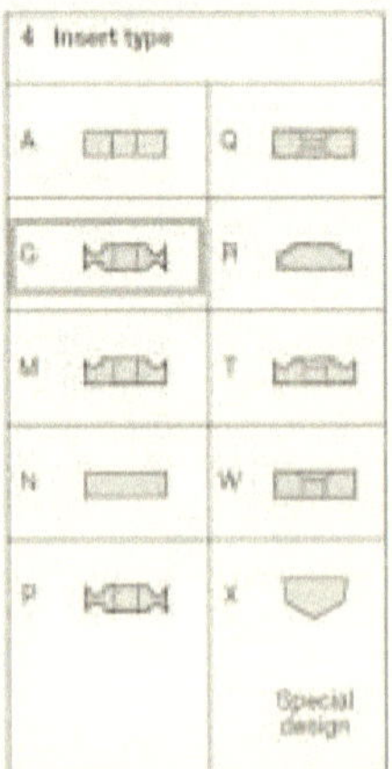

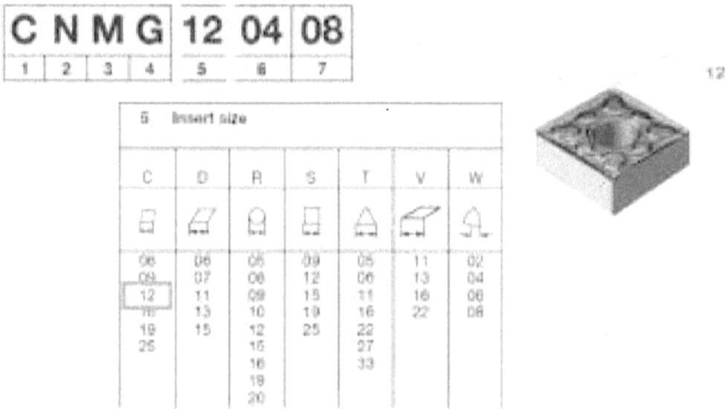
C N M G 12 04 08
1 2 3 4 5 6 7
5 Insert size
C D R S T V W
06 09 12 16 19 25
06 07 11 13 15
05 06 09 10 12 15 16 19 20
09 12 15 19 25
05 06 11 16 22 27 33
11 13 16 22
02 04 06 08
12

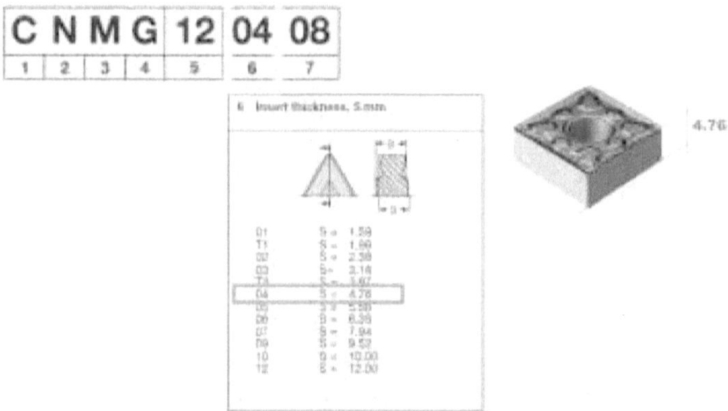
C N M G 12 04 08
1 2 3 4 5 6 7
6 Insert thickness, S mm
01 S = 1.59
T1 S = 1.98
02 S = 2.38
03 S = 3.18
T3 S = 3.97
04 S = 4.76
05 S = 5.56
06 S = 6.35
07 S = 7.94
09 S = 9.52
10 S = 10.00
12 S = 12.00
4.76

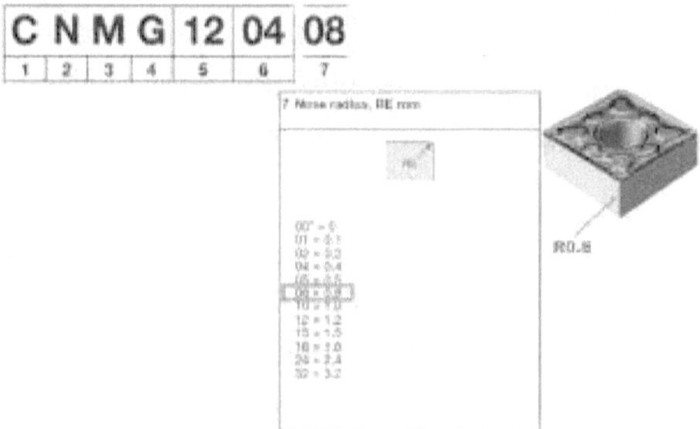
C N M G 12 04 08
1 2 3 4 5 6 7
7 Nose radius, RE mm
00* = 0
01 = 0.1
02 = 0.2
04 = 0.4
05 = 0.5
08 = 0.8
10 = 1.0
12 = 1.2
15 = 1.5
16 = 1.6
24 = 2.4
32 = 3.2
R0.8

5.3.3 Inserts comparison

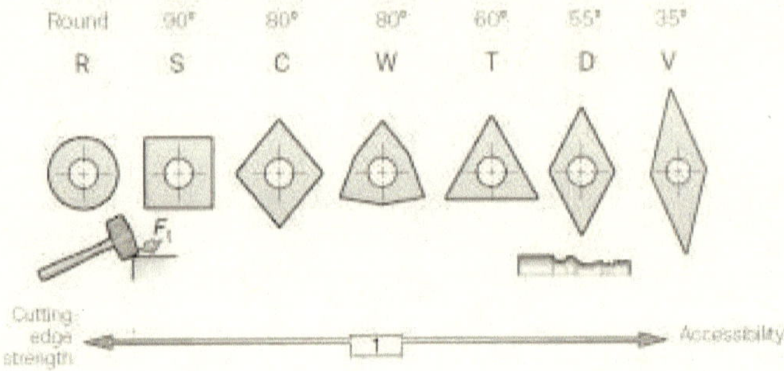

5.4 Cutting Tool Materials & Substrates

5.4.1 Solid Carbide Grades

- **Sub-Micro grain (0.2–0.5 µm):** Exhibit high wear resistance; preferred for hardened steels (>45 HRC) and precision finishing.
- **Micro grain (0.5–1.0 µm):** Offer higher toughness; handle interrupted cuts better; cost-effective for general machining.
- **Cobalt Content:**
    - Higher cobalt (8–12%) increases toughness, benefits interrupted cuts but lowers wear resistance.

- Lower cobalt (4–8%) creates harder but more brittle tools, better suited for abrasive materials.

5.4.2 ISO Material Groups

Group	Material	Characteristics
P	Steel	Moderate hardness, general-purpose
M	Stainless Steel	Sticky, work-hardening, heat prone
K	Cast Iron	Abrasive, brittle, often dry machining
N	Non-Ferrous	Soft, sticky, chip evacuation critical
S	Superalloys	Tough, heat-resistant, work-hardening
H	Hardened Steels	High hardness (>45 HRC), abrasive

5.4.3 Alternative Tool Materials

- **HSS/Co-HSS:** Tough and cost-friendly; limited to low-rigidity, low-speed applications.
- **Cermet/Ceramic:** Suitable for high-speed finishing in cast iron and superalloys; brittle, requiring stable setups.
- **PCD (Polycrystalline Diamond):** Excellent wear resistance with non-ferrous and abrasive materials like Al-Si and composites.
- **CBN (Cubic Boron Nitride):** Best for hardened steels (>50 HRC) with superior wear resistance.

5.5 Coating Selection Strategy

Coating	Best For	Avoid In	Benefits
TiN	General HSS, entry-level carbide	Aluminum finishing	Basic wear resistance
TiCN	Low/medium steel, cast iron	Aluminum, titanium	Harder than TiN, reduces friction

Coating	Best For	Avoid In	Benefits
AlTiN/ TiAlN	Steel, stainless, dry machining	Aluminum (risk of BUE)	Heat barrier; forms Al_2O_3 layer
AlCrN	Stainless, hardened steels	Soft aluminum	High hot hardness, oxidation resistance
ZrN/ TiB_2/ DLC	Aluminum, copper, plastics	Steel	Anti-stick, BUE prevention
Diamond (CVD/ PCD)	Graphite, CFRP, abrasives	Steel	Extreme wear resistance

5.6 Cutting Parameter Fundamentals

5.6.1 Formulas

- **Spindle Speed (RPM):**

$$N = \frac{1000 \times V_c}{\pi D}$$

Where Vc is surface speed (m/min), *D* is tool diameter (mm) and π =3.14

- **Feed Rate (mm/min):**

F=fz × z × n

fz is feed per tooth (mm/tooth), z is number of flutes.

5.6.2 Starting Parameters by Material

Material	***Vc*** **m/min**	***fz*** **mm/tooth**	**Flutes**	**Radial Engagement (*Ae*) %**	**Axial Engagement (*Ap*) (×D)**
Aluminum 6061-T6	250 – 450	0.03 – 0.10	3 – 4	40 – 60	1 – 2

Material	*Vc* m/min	*fz* mm/tooth	Flutes	Radial Engagement (*Ae*) %	Axial Engagement (*Ap*) (×D)
Mild Steel 1018/P20	120 – 220	0.02 – 0.06	4 – 6	30 – 50	0.5 – 1
4140 (30–35 HRC)	90 – 160	0.02 – 0.05	4 – 6	20 – 40	0.5 – 1
Stainless 304/316	70 – 140	0.01 – 0.04	5 – 7	10 – 20	1 – 2
Titanium Ti-6Al-4V	40 – 90	0.01 – 0.03	5 – 6	5 – 15	1 – 2
Cast Iron	150 – 250	0.02 – 0.06	4 – 6	30 – 50	0.5 – 1

Note: The values indicated here are only for reference. For actual parameters, refer cutting tool catalogue or get recommendations from cutting tool manufacturer.

5.7 Engagement Strategy

- **Conventional Roughing:** 40–60% radial, 0.5–1× axial; balances material removal and tool life.
- **High-Efficiency Machining (HEM):** 5–15% radial, 1–3× axial; optimizes chip thinning and productivity.
- **Finishing:** 3–8% radial, full axial wall depth pass for optimal surface finish.

5.8 Material-Specific Machining Strategies

- **Aluminum (6xxx/7xxx series):** Use 3–4 flute tools with sharp positive rake and ZrN/TiB_2/DLC coatings. Employ air blast or mist coolant to prevent built-up edge; high helix angles aid chip evacuation.
- **Plain Carbon/Alloy Steels:** Employ 4–6 flute tools with corner radius and AlTiN/AlCrN coatings. Use micro-honed edges and flood coolant for heat control.
- **Stainless Steel (304/316):** Utilize 5–7 flutes and high helix angles with AlCrN coatings. Maintain chip load carefully to avoid work hardening; use high-pressure coolant systems.

- **Titanium/Superalloys:** Select 5–6 flute tools with robust edges and AlTiN/AlCrN coatings. Avoid dwell time; apply high-pressure coolant.
- **Cast Iron:** Use 4–6 flutes with TiAlN or AlCrN coatings or uncoated tools. Prefer dry machining or minimal mist and employ dust collection for abrasive particles.

5.9 Tool Life Economics & Optimization

Taylor's Equation:

$Vc \times T^n = C$

Where Vc = cutting speed, T= tool life, and n, C are material constants.

Optimization Levers

- Increase chip load (feed per tooth *fz*) to reduce cycle time.
- Employ HEM strategies to improve efficiency.
- Reduce spindle speed *Vc* only if necessary to extend tool life.
- Strengthen edges with corner radius or micro-hone.
- Maintain tool runout ≤ 5–8 µm for finishing, ≤ 10–15 µm for roughing.

- Use hydraulic or shrink-fit holders to reduce vibration and extend tool life.

5.10 Troubleshooting Guide

Symptom	Likely Cause	Quick Fix
Built-Up Edge (BUE)	Wrong coating or low chip load	Switch to ZrN/TiB_2/DLC; increase fz; add air blast coolant
Flank Wear	High surface speed, abrasive material	Decrease Vc; switch to tougher carbide grade
Crater Wear	Excessive heat, poor coating	Use AlTiN/AlCrN; reduce Vc; increase fz
Edge Chipping	Rubbing, high runout, too sharp edge	Decrease runout; add micro-hone or radius; increase fz
Thermal Cracks	Overheating or coolant cycling	Ensure consistent coolant flow; reduce speed

Symptom	Likely Cause	Quick Fix
Notch Wear	Work-hardened surface	Rough cut to remove hardened layer; vary depth of cut
Chatter	Resonance, excessive tool overhang	Reduce stick-out; use variable helix tools; switch to hydraulic holders

5.11 Case Studies

- **6061 Aluminum Pocket:**
  10 mm, 3-flute, ZrN
  coating; Ae=20%, Ap=1.5×D, *Vc*=350 m/min, *fz*= 0.06 mm/tooth; 45-minute tool life; clean finish; no BUE.
- **4140 Steel (32 HRC):**
  10 mm, 6-flute, AlTiN coating, 0.5 mm corner radius; Ae=10%, *Ap*=2×*D*, *Vc*=140 m/min, *fz*=0.04 mm/tooth; consistent tool life, minimal wear.
- **304 Stainless Slot:**
  8 mm, 6-flute, high helix, AlCrN
  coating; Ae=12%, *Ap*=1.5×*D*, *Vc*=100 m/min; through-spindle coolant; no work hardening, smooth walls.

- **Titanium Ti-6Al-4V Thin Wall:**
  6 mm, 5-flute, AlTiN coating, robust edge; *Ae*=4%, full wall depth axial engagement; *Vc*=60 m/min; mirror finish, no chatter.

5.12 Quick Reference Checklists

5.12.1 Tool Selection

- Confirm material group (P, M, K, N, S, H).
- Select geometry (flute count, helix, edge prep).
- Choose coatings (e.g., ZrN for aluminum, AlCrN for stainless).
- Match corner profile (square, radius, ball) to the operation.
- Verify holder runout ≤ 5 µm for finishing.
- Limit tool overhang ≤ 4× diameter.

5.12.2 First Article Setup

- Define SFM and chip load by material.
- Document radial and axial engagement (Ae, Ap).
- Confirm coolant type and proper delivery.
- Simulate toolpaths in CAM software.
- Perform machine warm-up (10–15 minutes).
- Conduct tug testing on work holding.

5.12.3 Production Monitoring

- Inspect tool wear (flank, crater, BUE) on initial parts.
- Adjust chip load first, then SFM if required.
- Log stable parameters consistently.
- Monitor tool life and cost-per-part.
- Record QC outcomes for critical features.

5.13 Practical Exercises

- **Runout Impact:** Test 10 mm end mill in ER collet, hydraulic chuck, and high-precision holder; measure runout, finish, tool life on steel.
- **Endmill type Comparison:** Machine 1018 steel with square end and corner radius; record wear pattern, tool life, and economics.
- **Aluminum Coating Test:** Compare uncoated vs. TiB_2-coated end mill on 6061; evaluate BUE, finish, and productivity.

PRECISION RECAP

Mastering cutting tool geometry and grades drastically enhances machining outcomes. By selecting correct flute counts, helix angles, rake/relief angles, edge finishes, and coatings, and by applying suitable cutting parameters for each material, one can maximize tool life, surface quality, and cost-effectiveness.

Chapter 6: Fixturing Fundamentals — Stability, Accuracy & Repeatability

CNC FOCUS POINTS

This chapter covers:

- Core fixturing principles for stability, accuracy, and repeatability.
- Different fixture types and their applications.
- Dowel pin strategies for exact part location and thermal compensation.
- Clamping methods and optimization of clamping forces.
- Fixture design considerations focused on efficiency and ergonomics.
- Maintenance routines and common troubleshooting techniques.
- Advanced concepts and future trends in fixturing technology.

6.1 Introduction

A fixture is far more than a simple holder—it forms the backbone of precision CNC machining. Even the best programming and advanced cutting tools fail without a stable, accurate, and repeatable fixturing system. Poor fixturing causes dimensional inaccuracies, surface finish defects, accelerated tool wear, and safety risks. Conversely, a well-designed fixture enables lights-out manufacturing, reduces operator intervention, and ensures consistent high-quality parts.

6.2 The Role of Fixturing in CNC Manufacturing

Fixtures fulfill four essential functions:

- **Stability:** Prevent vibration and machining chatter by firmly resisting cutting forces.
- **Accuracy:** Position the workpiece precisely relative to the programmed coordinate system and cutting tool path.
- **Repeatability:** Enable consistent part placement, ensuring tolerances such as ±0.025 mm in general use and ±0.005 mm in high-precision operations.
- **Throughput:** Minimize setup time via quick-change mechanisms and automation, maximizing machine utilization.

- **Tool Life:** Minimize vibration and chatter to lengthen tool durability.

Key Insight: *A fixture acts as the silent partner bridging CNC programming and machining, ensuring the workpiece remains secure and properly aligned throughout the manufacturing process.*

6.3 Key Fixturing Principles

6.3.1 Locate Before You Clamp

Position the workpiece accurately using locators before applying clamping force. Avoid using clamps to force the part into position since this induces stress and misalignment.

In fixturing and CNC machining, a workpiece in space can move in **six Degrees of Freedom (6-DOF)** — three linear (translational) and three angular (rotational). To properly locate and clamp a part, we must **restrict all 6 degrees** so it cannot shift during machining.

6.3.2 The 6 Degrees of Freedom (DOF)

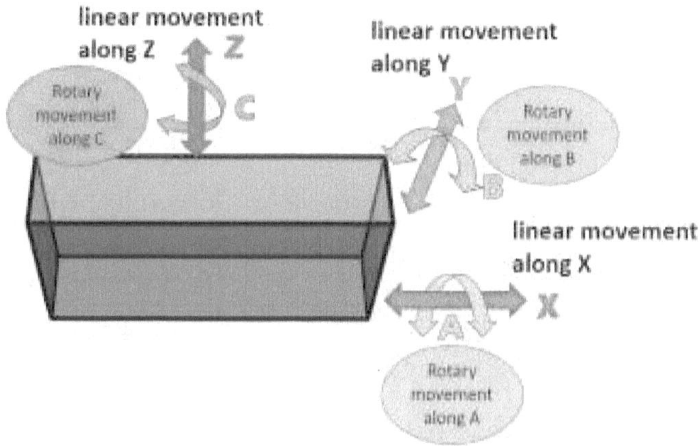

Type	Axis	Description	Example of Movement
Translational (Linear)	X	Sliding left–right	Part shifts sideways on the fixture
	Y	Sliding forward–back	Part moves toward or away from operator
	Z	Sliding up–down	Part lifts or sinks vertically

Type	Axis	Description	Example of Movement
Rotational (Angular)	Roll	Rotation about X-axis	Part tilts side-to-side
	Pitch	Rotation about Y-axis	Part nods up-down like a hinge
	Yaw	Rotation about Z-axis	Part rotates left–right like a compass dial

---

6.3.3 Why Is 6-DOF Important in Machining?

When placing a part in a fixture, you must **eliminate all 6 motions** using locators and clamps:

Motion Prevented By	Example
3-point support	Locator pads prevent vertical (Z) movement & roll/pitch
Side locators / pins	Prevent linear movement in X/Y directions

Motion Prevented By	Example
Dowel pins / stops	Prevent rotation (yaw)
Clamps	Hold the part down to maintain contact with locators

6.3.4 Design Philosophy: 3-2-1 Principle

A common fixturing method to control 6-DOF:

Surface Supports	DOF Restraint
3 points on primary surface	Stop Z, roll, pitch
2 points on secondary surface	Stop Y, yaw
1 point on tertiary surface	Stop X

Thus, **3-2-1 = 6 degrees restricted**

6.3.5 CNC Machining Example

Suppose you're machining a block:

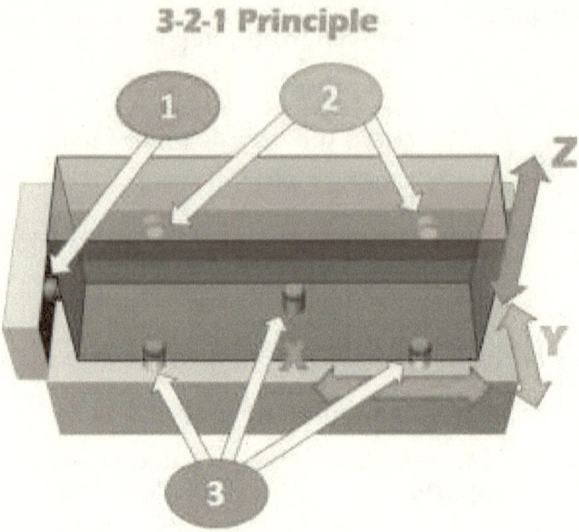

- It rests on **three pads** → No Z movement / no tilting
- **Two side pins** touch one face → Stops Y movement / yaw
- **One pin** touches the adjacent face → Stops X movement
- **Clamps** hold it down → Keep contact and prevent vibration

Now the part is fully constrained for accurate machining.

6.3.6 Simple Memory Tip

Move: X, Y, Z

Rotate: Roll, Pitch, Yaw

Total = 6

6.3.7 Why someone say 12 DOF?

The confusion sometimes comes from **counting both directions on each axis**:

- +X and –X
- +Y and –Y
- +Z and –Z
- +Rotation and –Rotation about each axis

That gives 12 directions — **but they are not 12 DOF**. They are **6 DOF with positive and negative directions**.

6.3.8 Minimize Overhang

Support the workpiece as close as possible to the cutting zone, as deflection increases approximately with the cube of unsupported length. For long or slender parts, use additional supports or steady rests.

6.3.9 Control Deflection

Design fixtures with stiffness 3 to 5 times greater than the workpiece. Optimize geometry, material choice, and support spacing to reduce flexure.

6.3.10 Thermal Stability

Match the coefficient of thermal expansion between fixture and workpiece materials to avoid distortion. Design heat dissipation paths such as coolant channels or thermal breaks. Use diamond profile dowel pins to accommodate thermal expansion while maintaining alignment.

6.4 Dowel Pins for Accurate Location

6.4.1 Types of dowel pin

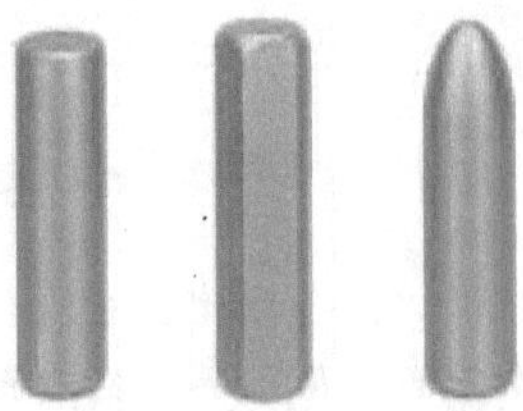

- **Parallel (Straight) Pins:** Restrain both X and Y axes; typically used on primary datum surfaces. Requires precision-reamed holes within H7 tolerance (±0.005 mm).
- **Diamond (Relief) Pins:** Restrain only one axis, permitting thermal expansion along the other, preventing over-constraint. Usually used paired with parallel pins.

- **Bullet (Tapered) Dowel Pins:** Facilitate quick loading; employed in secondary or quick-change setups.

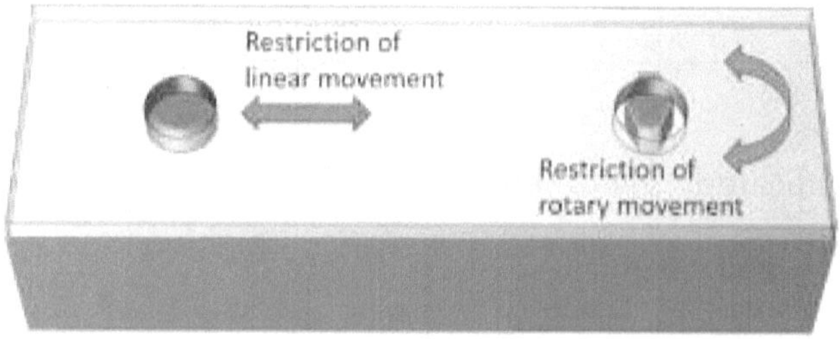

6.4.2 Best Practices

- Combine parallel and diamond dowel pins to maximize accuracy and accommodate thermal growth and or hole pitch variations.
- Use hardened tool steel pins with hardness Rc 58–62 for durability.
- Inspect pins regularly and replace if wear exceeds 0.005 mm.
- Ensure reamed holes are clean and burr-free before inserting pins.

6.5 Fixture Types and Applications

Fixture Type	**Strengths**	**Applications**
Dedicated Fixtures	Highest accuracy and repeatability; tailored to specific parts	Mass production, high-volume runs
Modular Fixtures	Flexible, adaptable, and reconfigurable	Prototyping, job shops, low-volume runs
Vises & Clamps	Cost-effective, versatile precision options	General-purpose small/medium parts
Magnetic Fixtures	Quick setup for ferrous parts; clamp-free	Thin and delicate parts, flat surfaces
Vacuum Fixtures	Hold thin, non-ferrous plates without mechanical stress	Sheet metal, composites, sealed surfaces

Fixture Type	Strengths	Applications
Soft Jaws	Custom-machined, repeatable gripping	Repeat setups, medium-volume production
Poka-Yoke Fixtures	Error-proofing using asymmetric locators or sensors	High-volume production, operator training

Pro Tip: *Use dedicated fixtures for long-term production and modular fixtures for adaptability in dynamic environments.*

6.6 Clamping Methods & Force Distribution

6.6.1 Clamping Types

- **Mechanical Clamps:** Strap, swing, or edge clamps; reliable but manual.
- **Pneumatic Systems:** Deliver 80–100 psi; fast, clean, suited to light-medium forces.

- **Hydraulic Systems:** Provide very high force (up to 10,000 psi); precise control; ideal for heavy machining, but require maintenance.
- **Toggle Clamps:** Ergonomic, quick, with visual engagement confirmation.

6.6.2 Force Distribution Principles

- Apply clamping force opposite cutting force direction to counteract vibration.
- Distribute force broadly to avoid localized deformation.
- Avoid unsupported clamping spans which cause bending.

6.6.3 Clamping Force Formula:

$$F_{clamp} = \frac{F_{cut} \cdot SF}{\mu}$$

where:

- Fclamp = Clamping force
- Fcut= Cutting force (from tool data or CAM simulations)
- SF= Safety factor (2–4 depending on material and operation)
- μ= Coefficient of friction (typically 0.1–0.3; steel-on-steel ~0.15)

Example: If Fcut=500 N, SF=3, and μ=0.2:

$$F_{clamp} = \frac{F_{cut} \cdot SF}{\mu} = \frac{500 \times 3}{0.2}$$

Fclamp = 7,500*N*

6.7 Fixture Materials

Material	Strengths	Applications
Steel	High stiffness and durability	Heavy-duty fixtures, long runs
Aluminum	Lightweight and corrosion resistant	Prototypes, light-duty setups
Cast Iron	Excellent vibration damping	Large fixtures, high-volume
Epoxy-Granite	Superior damping and thermal stability	Precision machining, metrology

Material	Strengths	Applications
Plastics/Composites	Non-marring (do not damage the part), lightweight	Delicate parts, specialized uses

Material Selection Tip: Match fixture materials to workpiece properties (thermal expansion, stiffness) and machining requirements.

6.8 Fixture Design for CNC Efficiency

- Maintain at least **12 mm (0.5 inch)** clearance for tools, probes, and tool changers—account for long-reach tooling.
- Design sloped surfaces (greater than 15°) and open zones to guide chips away from cutting areas.
- Include chip pockets or channels for high-volume machining environments.
- Provide unobstructed coolant access; accommodate both flood and through-spindle systems.
- Reduce setup times via dual-station fixtures (while one fixture is under machining, other one will be getting loaded).

- Apply quick-change or zero-point clamping systems for rapid setups.
- Standardize locating features for repeatability and interchangeability.

6.9 Case Study — Aerospace Bracket Fixture

Old Setup:

- Vise like fixture with soft jaws.
- Setup time: 15 minutes/part.
- Repeatability: ±0.05 mm.
- Throughput: 8 parts/shift.
- Issues: Slow setup, inconsistent clamping, high scrap rates.

New Fixture:

- Modular design with 3-2-1 locating (parallel + diamond pins), hydraulic clamps, holding 4 parts simultaneously.
- Setup time decreased to 2 minutes.
- Repeatability improved to ±0.005 mm.
- Throughput increased to 12 parts/shift.
- Scrap rate reduced by 90%.

- ROI: 3 months, driven by efficiency and quality gains.

Key Lesson: Investing in proper fixture design dramatically enhances efficiency, quality, and profitability.

6.10 Common Fixture Mistakes

Mistake	Consequence	Solution
Over-constraining	Stress, warping, misalignment	Use 3-2-1 principle; incorporate diamond pins
Excessive Clamping	Thin part deformation	Calculate minimal clamping force
Poor Ergonomics	Operator fatigue, slow setups	Design for accessibility; use ergonomic clamps
Lack of Maintenance	Worn pins/clamps,	Schedule regular

Mistake	Consequence	Solution
	reduced accuracy	inspections and upkeep

6.11 Fixture Maintenance & Inspection

6.11.1 Inspection frequency

- **Daily:** Inspect pins, clamps, and surfaces for wear, chips, or damage; clean thoroughly.
- **Weekly:** Deep clean with solvent, lubricate moving parts, calibrate clamp force.
- **Monthly:** Measure dowel pin wear; replace if >0.005 mm. Check fixture flatness and finish.
- **Yearly:** Conduct full dimensional inspection using Coordinate Measuring Machines (CMM) or precision gauges.

6.11.2 Troubleshooting Tips:

- Dimensional Drift: Often from worn pins, loose clamps, or thermal distortion. Inspect, re-torque, and adjust thermal compensation.
- Slow Setups: Caused by poor ergonomics or manual clamps; implement quick-change systems.

- Poor Surface Finish: Due to inadequate rigidity or chip buildup; increase fixture stiffness and improve chip evacuation.

6.12 Advanced Fixturing Concepts

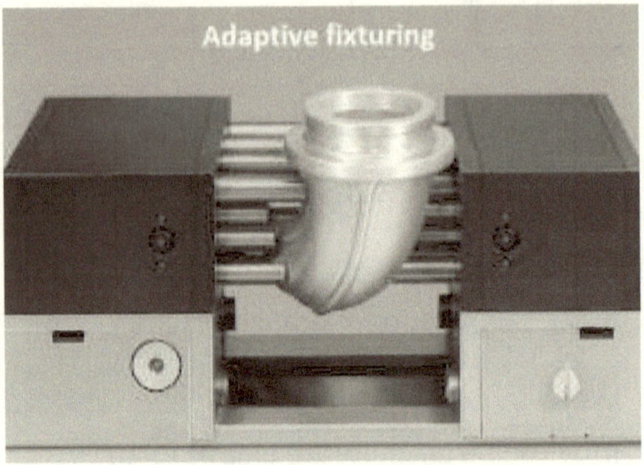

- **Automated Fixtures:** Employ robot loaders and sensors for part detection; integrate with Manufacturing Execution Systems (MES).
- **Adaptive Fixturing:** Use servo-driven locators and variable clamps adjusted dynamically via probe feedback.
- **Smart Fixtures:** Embed IoT devices monitoring load, vibration, and temperature in real-time.
- **Industry 4.0 Integration:** Use digital twins to simulate fixture performance before production; predictive maintenance using sensor data; cloud-based design optimization.

6.13 Economics of Fixturing

Costs

- Design and engineering (CAD, prototyping).
- Material and fabrication.
- Installation and operator training.

Benefits

- Reduced cycle times from faster setups.
- Improved part quality with lower scrap rates.
- Increased tool life due to better stability.
- Enhanced machine utilization enabling lights-out manufacturing.

6.14 Future Trends

- **Additive Manufacturing:** 3D-printed fixtures enable lightweight, conformal designs and rapid prototyping.
- **Advanced Materials:** Carbon fiber and nano-enhanced coatings improve durability and weight.
- **Digital Twins:** Virtual simulations optimize fixture design and predict stress points.
- **Smart Automation:** IoT and AI enable real-time fixture adjustments and predictive maintenance.

PRECISION RECAP

Fixtures underlie CNC machining success by ensuring stability, accuracy, and repeatability. Key points:

- Always locate before clamping, following the 3-2-1 principle to avoid over-constraint.
- Use parallel and diamond dowel pins to enhance precision while compensating for thermal expansion.
- Design fixtures with adequate tool clearance, chip evacuation paths, and coolant access.
- Select fixture materials based on stiffness, thermal properties, and application demands.
- Maintain fixtures proactively to prevent drift and downtime.
- Evaluate fixture economics carefully to justify investment via reduced cycle times and scrap.
- Adopt advanced technologies like automation, smart fixtures, and digital twins for future-ready manufacturing.

Remember: *Every detail—from base plate rigidity to dowel pin precision—must harmonize to deliver consistent, high-quality parts.*

Chapter 7: Choosing the Right Machine — Process Planning for Precision and Profitability

CNC FOCUS POINTS

This chapter provides Indian manufacturers a structured framework to align machine choices with part needs, operational goals, and financial realities. It compares different types of machines and guidance to choose the right machine.

7.1 Introduction

Selecting the right CNC machine is a critical decision that shapes a manufacturer's ability to deliver precision parts, meet delivery schedules, and optimize costs. Choosing a well-suited machine enhances competitiveness by improving efficiency, quality, and scalability. Poor choices lead to underutilized assets, production bottlenecks, increased costs, and lost market opportunities.

Modern machine selection requires evaluating technical specifications, automation capabilities, Industry 4.0 readiness, evolving customer requirements, and scalability projections.

7.2 The Strategic Role of Process Planning

Process planning ensures alignment between customer requirements and machine capabilities for efficient, high-quality production.

7.2.1 Process Planning Hierarchy

- **Part Analysis:** Assess geometry, material, tolerances, and inspection demands.
- **Operation Sequencing:** Plan machining steps, setups, tooling, and quality controls.
- **Resource Allocation:** Match machines to schedules, equip operators, and support infrastructure.

7.2.2 Critical Success Factors

- **Capability Matching:** Avoid undersized machines that miss specs or oversized costly buys.
- **Integration:** Ensure machines link smoothly with CAM, ERP/MES, and automation.
- **Scalability:** Buy with future technologies, materials, and volume growth in mind.

7.2.3 Benefits

- Reduce cycle time by 20–40%.
- Improve quality (Cpk) by 25–50%.

- Lower total cost of ownership through optimized resource use.
- Gain competitive edge with faster delivery and advanced capabilities.

7.3 CNC Machine Type Analysis (India-Adjusted Costs)

Disclaimer: The cost of machines indicated here are only for comparison purpose. Get the actual quotes from the machine manufacturer before making any buying decision.

7.3.1 3-Axis Vertical Machining Centers (VMCs)

- **Best For:** Simple prismatic parts, 2.5D machining, prototyping, operator training.
- **Capabilities:**
    - Work envelope: 300 mm × 200 mm × 200 mm to 2000 mm × 1000 mm × 760 mm
    - Spindle speed: 8,000 – 40,000 RPM
    - Accuracy: ±0.005 mm to ±0.02 mm
- **Advantages:** Budget-friendly, easy programming, mature technology.
- **Limitations:** Requires multiple setups for complex geometries; limited angled access.

- **Economics:** ₹30 lakhs to ₹1.6 crores (₹3 million to ₹16 million) (~$37,000 to $195,000)
- **Payback Period:** 2–4 years typical in India.

7.3.2 4-Axis Horizontal Machining Centers (HMCs)

- **Best For:** Automotive, heavy cavities, multi-side machining with superior chip removal.
- **Capabilities:** Rotational axis ±0.001° precision; spindles 15 to 100 HP; pallet changers for continuous running.
- **Advantages:** Heavy-duty cutting, automation ready with better ergonomics.
- **Limitations:** Larger footprint, more complex operation, and maintenance.
- **Economics:** ₹1.2 crores to ₹4.5 crores (~$150,000 to $570,000)
- **Payback Period:** 3–6 years.

7.3.3 5-Axis Machining Centers

- **Styles:** Trunnion table or swivel head configurations.
- **Best For:** Aerospace parts, medical components, molds, complex prototypes.
- **Advantages:** Single-setup capabilities, improved accuracy and surface finish, longer tool life.

- **Limitations:** High cost (2–4× VMC), advanced programming complexity.
- **Economics:** ₹1.6 crores to ₹9.5 crores (~$210,000 to $1.2 million)
- **Payback Period:** 4–8 years.

7.3.4 CNC Turning Centers and Lathes

- **Types:** Horizontal for shafts and rods, vertical for large flat components.
- **Advantages:** Superior concentricity, high material removal rates, fine finish.
- **Limitations:** Limited to rotational parts; basic milling on some models.
- **Economics:** ₹15 lakhs to ₹3.75 crores (~$19,000 to $480,000)
- **Payback:** Varies depending on use case.

7.3.5 Mill-Turn Centers

- **Capabilities:** Combine turning, milling, Y-axis, sub-spindle, and automation for complex machining.
- **Best For:** Medical implants, aerospace, and automotive transmission parts.
- **Advantages:** Single-operation machining reduces handling; high positional accuracy.

- **Limitations:** Capital intensive; demand highly skilled operators.
- **Economics:** ₹4 crores to ₹22 crores (~$510,000 to $2.8 million)
- **Payback Period:** 5–10 years.

7.4 Selection Criteria

7.4.1 Part Geometry & Complexity

- **Simple (3-Axis):** Orthogonal features, ±0.125 mm tolerances.
- **Moderate (4-Axis):** Angled features, contoured surfaces, multi-side access.
- **Complex (5-Axis):** Sculptured surfaces, compound angles, and tight tolerances.

7.4.2 Materials

- **Easy:** Aluminum, brass; high spindle speeds (15,000–40,000 RPM).
- **Moderate:** Steel, cast iron; rigid frames, speeds 3,000–15,000 RPM.
- **Hard:** Titanium, Inconel; low speeds 1,000–8,000 RPM, high torque, through-spindle coolant.

7.4.3 Production Volume

- **Low (1–100 parts):** Flexible 3-axis for short runs and prototypes.
- **Medium (100–5,000 parts):** Hybrid flexibility; semi-automated (4-axis HMCs).
- **High (5,000+ parts):** Full automation, Statistical Process Control, lights-out machining.

7.4.4 Automation Levels

- Level 1: Manual loading (prototypes).
- Level 2: Semi-automatic tool/pallet changers.
- Level 3: Fully automated with robots and in-process inspection.

7.4.5 Work Envelope & Space Planning

Calculate travel: Part size + fixture + tool clearance + 10% margin. Include considerations for floor space, material flow, utilities, and servicing.

7.4.6 Spindle Specifications

Material	Speed (RPM)	Torque	Notes
Aluminum	15,000–40,000	Moderate	High speed, moderate torque

Material	Speed (RPM)	Torque	Notes
Steel	3,000–15,000	High	Rigid bearings, heavy load
Titanium	1,000–8,000	High	Robust bearings, coolant flow

5.4.7 Control Systems

- **Fanuc:** Widely used, dependable, best for automotive workflows.
- **Siemens:** Advanced multi-axis control, strong Industry 4.0 connectivity.
- **Heidenhain:** High-precision, aerospace & mold applications; conversational programming easing complex jobs.

7.5 Cost Components for ROI

- Capital: Machine price, GST (~18%), installation, tooling, and training.
- Operating: Labor (generally lower costs in India), energy, maintenance, consumables.

Capacity Planning

Calculate productive hours × utilization rate (70–85%) × efficiency (75–90%).

7.6 Case Study — Aerospace Component Selection

- **Company:** MidWest Aerospace Solutions (Titanium brackets)
- **Old setup:** 6–8 setups on 3-axis VMCs, 8.5-hour cycle, 12% scrap, 8 parts/shift.
- **Options:**
    - 3-Axis VMC – Low cost but inadequate for geometry.
    - 4-Axis HMC – Reduced setups but still multiple.
    - 5-Axis VMC – Single setup, 65% cycle time reduction.
    - 5-Axis HMC – Premium automation, higher cost.
- **Chosen:** DMG Mori NVX5100 5-axis VMC (~₹5.6 crores / $703,000).
- **Results:** Cycle time to 2.8 hours, scrap down to 1.8%, revenue up ₹33 crore (~$4.2 million) annually, ROI in 18 months.

- **Lessons:** Intensive operator training (160 hours), tooling and fixturing optimization, strong supplier collaboration.

7.7 Common Mistakes

- Underestimating future capacity.
- Over-investing in machines with unused advanced features.
- Deploying complex machines without adequately trained staff.
- Ignoring integration with software and utilities.
- Focusing narrowly on price rather than TCO (Total cost of Ownership)

7.8 Advanced Considerations

- Industry 4.0: Predictive maintenance, digital twins, Overall Equipment Effectiveness (OEE).
- Sustainability: Energy-efficient machines, coolant recycling, chip reuse.
- Cybersecurity: Access controls, encryption, backups.
- Workforce: Operator training and ergonomic machine design.

7.9 Future Trends

- AI & machine learning for adaptive machining and predictive analytics.
- Hybrid additive-subtractive machines for molds and aerospace.
- Modular, scalable, and reconfigurable CNC systems.
- Nano-scale micro-machining for medical and electronics.

7.10 Selection Framework

Multi-Criteria Decision Analysis (MCDA)

Example weights: Technical 35%, Economic 30%, Strategic 20%, Risk 15%

Machine	Technical (35)	Economic (30)	Strategic (20)	Risk (15)	Total Score
3-Axis VMC	80	90	70	85	82.5
4-Axis HMC	85	75	80	70	79.5

Machine	Technical (35)	Economic (30)	Strategic (20)	Risk (15)	Total Score
5-Axis VMC	95	65	90	60	80.5

Risk Matrix

Risk	Likelihood	Impact	Mitigation
Obsolescence	Medium	High	Choose upgradeable/modular machines
Skill Gaps	High	Medium	Invest in comprehensive training
Demand Shifts	Medium	Medium	Diversify product portfolio

7.11 Implementation Best Practices

- **Phased Approach:** Define specs and volume, select machine based on MCDA and risk, install with vendor support, train operators, pilot production.
- **KPIs:** Compliance with tolerances, Cpk >1.33, OEE >85%, scrap <2%, ROI within target periods (KPI- Key Performance Indicator, Cpk – Process Capability Index, OEE- Overall Equipment Effectiveness, ROI- Return on Investment).
- **Continuous Improvement:** Regularly upgrade tooling, fixtures, parameters; monitor bottlenecks; plan control system updates.

PRECISION RECAP

Smart machine selection determines precision, profitability, and competitive positioning. Start with process-driven requirements, evaluate total costs over 5–7 years, prioritize scalable and Industry 4.0-ready systems, and confirm software and skill compatibility. Develop strong supplier partnerships for support and training. The right machine is not only a production asset but a foundation for future growth and success in precision CNC manufacturing.

Chapter 8: Machining Strategies for Accuracy, Speed & Tool Life

CNC FOCUS POINTS

This chapter covers roughing and finishing strategies of machining, high-efficiency milling (HEM), deflection control, material-specific tactics, advanced trends, and practical recommendations for manufacturing excellence.

8.1 Introduction

Success in modern CNC machining hinges on balancing three interconnected pillars: **accuracy**, **speed**, and **tool life**. Optimizing one factor often requires a trade-off with the others. The best machining strategies integrate deliberate choices in toolpaths, cutting parameters, tool selection, and continuous process optimization to:

- Increase productivity by reducing cycle times
- Meet or exceed tolerance and surface finish standards
- Lower cost per part by maximizing tool life and minimizing scrap

8.2 Roughing vs. Finishing Operations

8.2.1 Roughing Operations

Goal: Remove material quickly while leaving sufficient stock for finishing, maximizing Material Removal Rate (MRR).

Toolpath Strategies

- *Adaptive Clearing:* Maintains constant tool load, suitable for complex shapes.
- *Pocket Roughing:* Uses optimized entry/exit moves, such as helical plunging.
- *Zigzag/Spiral:* Ideal for large areas, reduces air cutting.
- *Rest Roughing:* Focuses on uncut material left by larger tools to even out stock for finishing.

Cutting Parameters

- Feed Rate: 2–5× higher than finishing (0.05–0.15 mm/tooth)
- Depth of Cut: 50–80% of tool diameter
- Spindle Speed: 60–80% of max RPM
- Stepover: 40–70% of tool diameter
- Tools: Robust corners (corner-radius end mills), carbide or coated (TiAlN, AlCrN), large diameters for stability

Stock Allowance Recommendations

- General: 0.25–0.76 mm
- Precision: 0.13–0.38 mm
- Hard materials (>45 HRC): 0.08–0.25 mm

8.2.2 Finishing Operations

Goal: Achieve specified tolerances and surface finish (typically Ra 0.4 to 1.6)

Toolpath Strategies

- *Parallel Finish:* Uniform passes for flat/curved surfaces
- *Constant Scallop Height:* Uniform surface across 3D geometries
- *3D Contouring:* For molds/aerospace parts

Parameters & Tools

- Feed Rate: 0.025–0.125 mm/tooth
- Depth of Cut: 0.25–1.2 mm
- Spindle Speed: 80–100% of max RPM
- Stepover: 5–20% of tool diameter
- Tools: High-helix, sharp edges for clean finish; ball end mills for contours; advanced coatings like ZrN, DLC

8.3 High-Efficiency Milling (HEM) & Advanced Roughing

8.3.1 Fundamentals

- Use low radial engagement (5–20% diameter) for consistent chip thinning
- High axial depth (up to flute length or 2× diameter)
- Faster feeds (0.1–0.3 mm/tooth)
- Material-optimized RPM (e.g., 10,000–20,000 for aluminum)

HEM Advantages

- Tool life extended 200–400% due to reduced forces
- Cycle times cut by 20–50%
- Lower heat, better chip formation

8.3.2 Trochoidal & Adaptive Milling

- *Trochoidal Milling*: Circular paths keep tool load steady, great for tough alloys
- *Adaptive Milling*: CAM software maintain optimal loads; ideal for variable geometries and complex parts

8.4 Step-Down & Step-Over Optimization

Axial Depth of Cut (Step-Down)

- Rigid setups: 1× diameter
- Standard: 0.5–0.8× diameter
- Long-reach: 0.2–0.5× diameter
- Finishing: 0.25–1.2 mm

Radial Step-Over Recommendations

Material	Step-Over (% of Diameter)
Aluminum	55–70%
Steel	40–60%
Stainless Steel	30–45%
Titanium	20–40%

Note: Use <45% or >55% to maintain stable cutting forces and surface finish

Tip: *For 3D finishing with ball end mills, use smaller step-overs for the smoothest contours.*

8.5 Deflection Control

Tool Deflection

Minimization Strategies

- Use the largest tool diameter practical
- Keep L/D (length/diameter) ratio <4:1
- Opt for stub-length tools and rigid holders (shrink-fit/hydraulic)
- Use climb milling and variable-helix tools

Workpiece Deflection

Minimization Strategies

- Maximize fixture support near cutting zones
- Use symmetrical material removal to balance stresses
- Employ semi-finishing passes and consider temporary supports

8.6 Climb vs. Conventional Milling

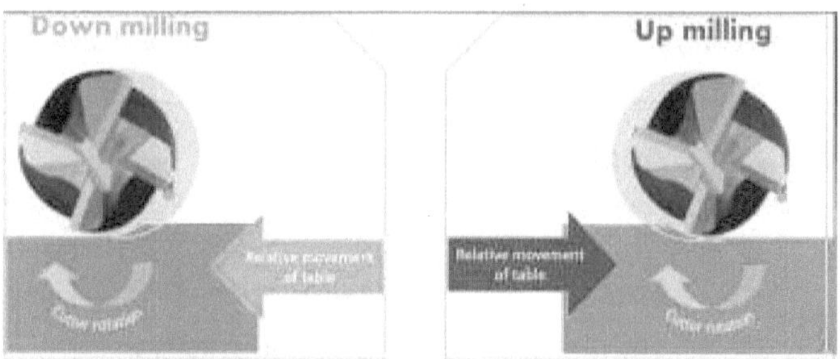

- **Climb Milling (Down Milling):** Softer finish, less tool wear, lower forces; requires rigid setup
- **Conventional Milling (Up Milling):** Useful for tough surfaces or older machines (backlash); higher forces, more wear

Best Practice: *Prefer climb milling for most CNC work unless facing hard layers or compromised rigidity.*

8.7 Sequence Optimization

- **By Feature:** Machine datums/case surfaces first, then functional features, then complex geometries
- **By Tool:** Group tool operations to minimize changes
- **By Heat Management:** Alternate heavy/light passes; schedule cool-downs for demanding materials
- **By Setup:** Minimize fixture and WCS changes

8.8 Coolant & Chip Management

Coolant Types & Their Best Applications

Type	Application	Notes
Flood	Steel, stainless	Maximum cooling/lubrication

Type	Application	Notes
High-Pressure	Deep holes, tough alloys	300–1000 PSI, through-spindle
Mist	Aluminum, finishing	Low residue
Air Blast	Dry machining, chip removal	Prevents buildup, BUE

- Plan toolpaths for chip evacuation (trochoidal/adaptive paths)
- Employ conveyors, blasts, and fixture design (open channels, slopes) to control chip flow

8.9 Material-Specific Strategies

Material	Strategy
Aluminum	High RPM, dry/air blast, aggressive feeds, ZrN/DLC coating
Stainless Steel	Flood/high-pressure coolant, positive rake tools, AlCrN coating

Material	Strategy
Titanium	Low RPM, high-pressure coolant, robust tools, frequent tool changes because of less tool life
Hardened Steel	CBN/ceramic tools, light cuts, low feeds, dry/mist coolant

8.10 Case Studies

Aerospace Titanium Bracket

- Old: Conventional roughing, 18.5 hr cycle
- New: HEM, 5-flute AlTiN-coated, trochoidal path — 13.5 hr cycle, 40% more tool life, $145 savings/part

Medical Device Implant

- Multi-stage finishing with carbide, constant scallop strategy
- Achieved Ra 4 µin finish, 22% faster cycle, zero scrap

8.11 Common Pitfalls

Pitfall	Result	Solution
Ignoring Tool Wear	Poor finish, dimensional errors	Monitor wear, adjust parameters, schedule changes
Excessive Deflection	Chatter, tool breakage	Reduce overhang, use larger/rigid tools/holders
Poor Parameter Setup	Long cycles, tool damage	Match roughing/finishing settings to material/goals

8.12 Future Trends

- **Adaptive Control:** Real-time feedback on spindle/vibration for auto adjustment
- **AI Optimization:** Machine learning selects best parameters and toolpaths
- **Advanced Materials:** Nano-coatings, hybrid ceramics (SiAlON)

- **Digital Twins:** Simulate strategies to predict tool life/surface quality
- **Industry 4.0:** IoT sensors for live monitoring of tool wear, deflection

PRECISION RECAP

Mastering machining strategies means continually balancing accuracy, speed, and tool life. Core practices include:

- Tailoring roughing and finishing parameters to each job's needs
- Integrating HEM/trochoidal paths for efficiency
- Controlling tool and workpiece deflection rigorously
- Adapting strategies to material and production volumes
- Managing coolant and chip removal actively
- Monitoring tool condition and refining processes based on performance data

Final Note: *Winning CNC manufacturing is a cycle of planning, execution, monitoring, and improvement. By leveraging data, new tech, and proven techniques, shops maximize precision, productivity, and profits.*

Chapter 9: Workshop Calculations Every CNC Machinist Should Know

> CNC FOCUS POINTS
>
> In this chapter, you will learn how to apply essential workshop calculations that directly impact CNC machining performance and accuracy. You'll understand how to determine feeds, speeds, chip load, and material removal rates to balance tool life and productivity. You will also develop the ability to select accurate tap drill sizes, convert between metric and imperial units, and apply trigonometry to fixturing and inspection tasks. Additionally, you will learn techniques to estimate machining time, horsepower, and torque requirements for real-world machining scenarios. By mastering these calculations, you will improve precision, reduce trial-and-error, and enhance your confidence on the shop floor. Ultimately, this chapter will help you use mathematics as a powerful tool to solve machining challenges and optimize CNC performance.

9.1 Introduction

Mathematics in CNC machining is not theoretical—it's a crucial, hands-on tool that empowers machinists to achieve precision, solve shop-floor challenges, and optimize efficiency. Mastery of workshop calculations helps CNC professionals. Solid command of these calculations reduces cycle times, maximizes tool life, boosts accuracy, and sharpens troubleshooting skills.

9.2 Feed Rate Calculation

Formula:

$F = N \times fz \times Z$

- F: Feed rate (mm/min or in/min)
- N: Spindle speed (RPM)
- fz: Feed per tooth (mm/tooth or in/tooth)
- Z: Number of flutes

Example:
4-flute end mill,
aluminum, N=2000RPM, fz=0.05 mm/tooth:

$F = 2000 \times 0.05 \times 4 = 400$ mm/min

Tips:

- For harder materials, decrease fz.
- Use higher fz with coated tools.
- Lower feed rates with less rigid setups to avoid vibration and chatter.

9.3 Spindle Speed Calculation

Formula:

$$N = \frac{1000 \times V_c}{\pi D}$$

- N: RPM
- Vc: Cutting speed (m/min)
- D: Tool diameter (mm)

Example:
10 mm tool, steel, *Vc*=150m/min:

N=(1000×150)/(3.1416×10)≈4775 RPM

Notes:

- Confirm the spindle speed does not exceed machine limits.
- Use manufacturer recommendations for cutting speeds by material.

9.4 Chip Load Analysis

Formula:

fz=F/(N×Z)

Typical Chip Loads:

Material	Chip Load (mm/tooth)
Aluminum	0.05–0.15
Steel	0.02–0.08
Stainless Steel	0.03–0.10
Titanium	0.01–0.05

Best Practices:

- Too high: Breakage, poor finish.
- Too low: Rubbing, rapid wear.
- Monitor chip color—silver for aluminum, blue for steel—short, curled chips signal efficiency.

9.5 Material Removal Rate (MRR)

Formula (milling):

$MRR = W \times D \times F$

- W: Width of cut
- D: Depth of cut
- F: Feed rate

Example:

W=5 mm, *D*=2 mm, *F*=400 mm/min:

MRR=5×2×400=4000 mm3/min

Uses:

- Estimate cycle time (Time=Volume/MRR)
- Compare tool/process productivity
- Check if the machine has adequate power

9.6 Horsepower & Torque

Horsepower Formula (Metric units):

$$kW = \frac{MRR \times K_s}{60 \times 10^3}$$

Where:

- MRR = Material Removal Rate (mm³/min)
- Ks = Specific Cutting Force (N/mm²) (also called specific energy or unit cutting force)

Horsepower Formula (Imperial units):

$$HP = \frac{MRR \times Unit\ Cutting\ Force}{33000}$$

Where:

- **MRR** = Material Removal Rate (in³/min)

- **Unit Cutting Force (UCF)** = Specific cutting pressure (psi) for the material
- **33000** = ft-lb/min per HP converted to in-lb/min (550 × 60)

Thumb rule

$$kW \approx \frac{MRR(mm^3/min)}{30,000 \text{ for steel}}$$

$$HP \approx \frac{MRR(in^3/min)}{2 \text{ for steel}}$$

Material	Unit Power (kW/cm³/min)
Aluminum	0.3–0.7
Steel	1.5–3.0
Stainless Steel	2.0–4.0
Titanium	3.0–6.0

Example:

MRR=4000 mm³/min, steel (U=2.0):

HP=4000×2.0/(60×1000)=0.133 kW

Torque Formula in metric:

$$T(Nm) = \frac{kW \times 9550}{RPM}$$

Where:

- T = Torque (in **ft·lb** for imperial & **Nm** for Metric)
- Power **HP** for imperial and **KW** for Metric
- **RPM** = Spindle speed

Torque Formula in imperial:

$$T(ft \cdot lb) = \frac{HP \times 5252}{RPM}$$

Tip: Always apply a safety factor (1.5x–2x) for load spikes and machine losses.

9.7 Trigonometry for Machinists

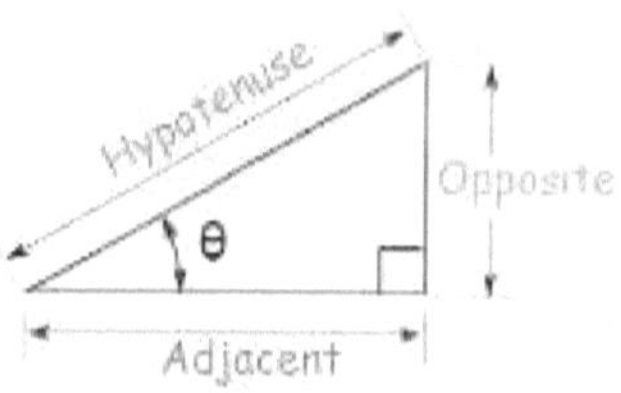

$$\sin(\theta) = \frac{Opposite}{Hypotenuse}$$

$$\cos(\theta) = \frac{Adjacent}{Hypotenuse}$$

$$\tan(\theta) = \frac{Opposite}{Adjacent}$$

9.8 Pythagoras Theorem

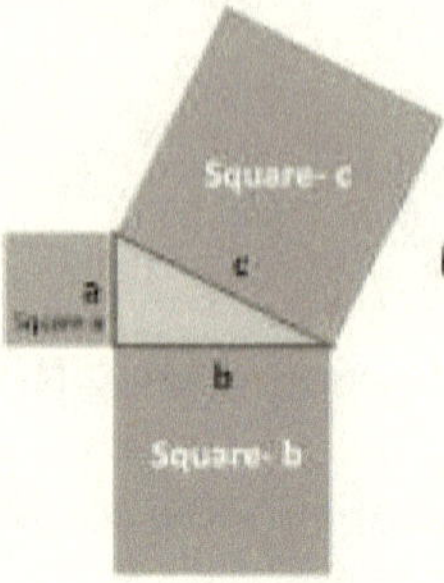

Pythagoras theorem

$$a^2 + b^2 = c^2$$

9.9 Formula for circle

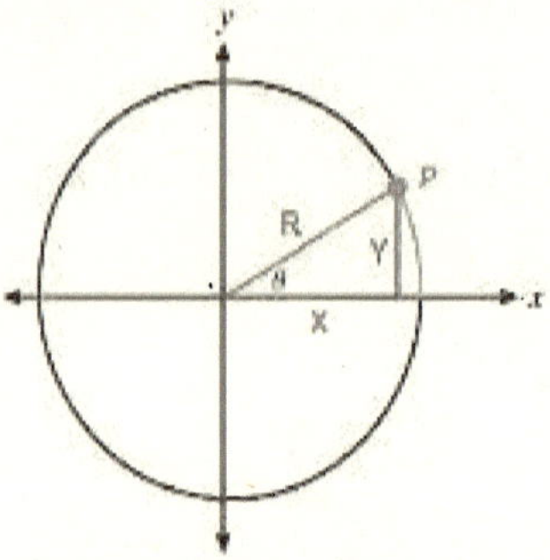

$$R^2 = x^2 + y^2$$

$$X = R \cos \theta$$

$$Y = R \sin \theta$$

9.10 Taper Angle Calculation

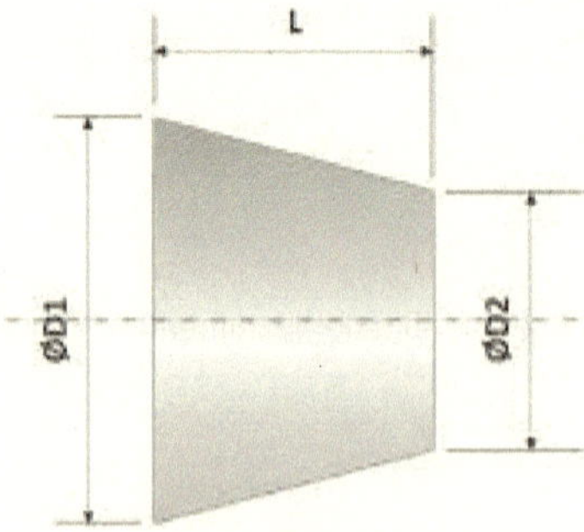

$$\tan\theta = \frac{D_1 - D_2}{2L}$$

9.11 Applications of calculation

- **Chamfer Angle:** Find angle for a 2 mm x 8 mm chamfer:
  $\theta = \tan^{-1}(2/8) = 14.04^\circ$

- **Bolt Circles:** For radius R, angle θ:
  $X = R\cos\theta,\ Y = R\sin\theta$

- **Taper Angle:**
  $\tan\theta = (D1 - D2)/2L$

9.12 Tap Drill Size

Formula:

Tap Drill=Major Diameter−Pitch

- For 75% thread engagement (general purpose)
- Check tap manufacturer's chart for optimal engagement

Examples:

- M10 × 1.5 → 10−1.5=8.5 mm drill
- 1/4"-20 UNC → 0.250−(1/20)=0.200 in → #7 drill

9.13 Unit Conversions

Conversion	Formula
in ↔ mm	1 in = 25.4 mm

Conversion	Formula
ft ↔ mm	1 ft = 304.8 mm
m/min ↔ SFM	1 m/min ≈ 3.281 SFM

Quick Rules:

- mm → in: divide by 25.4
- m/min → SFM: multiply by 3.281

9.14 Case Study: Feed Rate Optimization

Setup: 4-flute, 8 mm carbide end mill, 6061 aluminum
Initial: N=12,000 RPM, fz=0.02 → F=960 mm/min
Optimized: fz=0.04 → F=1920 mm/min

- Cycle time cut 50%
- Surface finish improved, tool life maintained
- Productivity up 18%

Lesson: Double the chip load, double the output—without sacrificing quality, provided calculations and monitoring are in place.

9.15 Common Mistakes & Solutions

Mistake	Consequence	Solution
Feed miscalculated	Excess wear/chatter	Use correct flute count, check runout
Wrong speed	Burned/broken tools	Adjust speed for tool, diameter, setup
Unit confusion	Scrap/crashes	Double-check units
Ignore geometry	Poor finish	Factor in deflection, thermal changes

9.16 Advanced Calculations

- **Force per tooth**

$$F_c = K_c \cdot f \cdot a_p$$

For milling, multiply further by number of teeth to get total force

- **Surface Roughness feed mark effect (milling):**

$$R_a \approx \frac{f_z^2}{32R}$$

Where R is the endmill corner radius.

- **Surface Roughness stepover effect (milling):**

$$h = R - \sqrt{R^2 - \left(\frac{s}{2}\right)^2}$$

Where R is the radius of ball cutter.

For small stepovers, formula can be modified as below;

$$h \approx \frac{s^2}{8R}$$

- **Surface Roughness (turning):**

$$R_a \approx \frac{f^2}{32r}$$

Where f is the feed rate in mm/rev & r is the tool nose radius

Notes

- Formulas assume perfect tool geometry and no vibration.
- Real finish (surface roughness) affected by:
  - Tool wear & nose radius
  - Tool material & coating

- Workpiece hardness
- Coolant & chip evacuation
- Machine rigidity & bearings
- Balanced tool setup
- Cutting parameters (speed/feed/depth)

9.17 Digital Tools & Verification

- Use manufacturer calculators (Sandvik, Kennametal, Guhring)
- CAM software often includes real-time parameter optimization
- Mobile apps: Machinist Calc Pro, Machinist Handbook Online
- Verify results: Inspect chips, listen for chatter, measure finish, compare to trusted data

PRECISION RECAP

Workshop calculations are the backbone of efficient, precise CNC machining. Core masteries for machinists include:

- Calculating optimal feeds and speeds tailored to job and setup
- Managing chip load and MRR for productivity and tool health

Success Factors:

- Start with data-driven parameters and adjust using feedback from machine, chips, and part condition.
- Document calculations for consistency and future reference.
- Train to identify process issues from sounds, chips, and tool wear.
- Leverage digital and CAM tools to excel.

Final Note: *Calculations are the starting line—mastery comes from interpreting the results and continuously refining based on real-world feedback. This is what sets a skilled CNC machinist apart as a true shop-floor problem solver and process engineer.*

Chapter10 - Macro Programming (Custom Macros / Parametric Programming)

CNC FOCUS POINTS

In this chapter, you will learn how macro programming transforms CNC code from fixed instructions into flexible, intelligent logic. You'll understand how to use variables, formulas, loops, and conditional checks to automate repetitive tasks and program families of parts without rewriting code. You will explore real applications such as bolt-circle drilling, slot machining, and trigonometric coordinate calculations. You will also learn to build error-proof programs using system variables, custom alarms, and input validation. By the end, you'll be able to create smart, reusable CNC routines that boost productivity, reduce programming time, and bring engineering-level control to your machining processes.

Usage of macro programming without proper understanding can be disastrous as it writes the information in to the system. Learn it fully and use it with extreme care.

10.1 Introduction

Macro programming lets you turn repeated machining logic into reusable, intelligent code. Instead of hardcoding numbers, you can use variables, formulas, logic, loops, and even trigonometry inside CNC programs. This enables:

- Automated calculations (toolpaths, depth steps, hole coordinates, etc.)
- Parametric programs (same program runs for different part sizes)
- Error-proofing and conditional checks
- "Smart cycles" for features your control doesn't natively support

You will see this called:

- Custom Macro, Custom Macro B (Fanuc)
- Parametric programming
- User macro / subprogram with arguments

10.2 Macro Variables

In CNC macro programming, variables behave just like variables in mathematics: they act as placeholders that store values that can change based on the situation. Just as algebra uses symbols like *x* and *y* to represent numbers in equations, CNC uses variables like #1, #100, or system variables to represent coordinates, dimensions, offsets, or calculated values. For example, area of a circle is defined as **A= πr²** where r is the radius of the circle. Here, r is the variable. By assigning the radius value of different circles, you can calculate the area of any circle. Same way, by assigning values and performing arithmetic operations on variables, you can make programs flexible and generic—capable of adjusting to different part sizes or machining conditions without rewriting the code. This connection to mathematical logic helps machinists automate decision-making, reuse code, and build "smart" machining routines that adapt just like formulas do in algebra.

10.2.1 Local Variables #1 – #33

- Scope: valid only inside the active macro call.
- Lifetime: cleared when the macro ends (like local variables in a subroutine).
- Typical use: arguments to subroutines.

Example:

- If you call:
  G65 P9000 A10.0 B25.0
  then inside O9000 you'll read:
    - #1 = 10.0 (A → #1)
    - #2 = 25.0 (B → #2)

Common mapping (Fanuc convention):

- A → #1

- B → #2
- C → #3
- D → #7
- E → #8
- F → #9
- H → #11
- I → #4
- J → #5
- K → #6
- R → #18
- etc.
  Note: The exact mapping can vary by control/option, but this is the common standard in Fanuc-type controls.

10.2.2 Common (Global) Variables #100 – #199 and #500 – #999

- Scope: controller-wide. You can read/write them from any program.
- Lifetime:
    - #100–#199 usually volatile (cleared on power-off).

    - #500–#999 usually non-volatile (kept after power-off).
- Use cases:
    - Sharing data between programs
    - Storing job settings (e.g. workpiece width, thickness, pitch, etc.)
    - Counters and flags (e.g. "did we probe already?")

Example:

#500 = 42.0 (store master part thickness)

#101 = #101 + 1 (increment cycle counter)

10.2.3 System Variables

- Reserved variables the control uses for machine state (spindle speed, axis positions, alarms, etc.).
- Read-only or read/write depending on the variable.
- Range and meaning are machine- and builder-specific.
- Examples:
    - #5021, #5022, #5023 → current machine position X/Y/Z
    - #5061–#5063 → probed position X/Y/Z (on probing-equipped controls)

These are extremely powerful for automation, in-process inspection, and adaptive machining. Use with caution.

10.3 Mathematical Operations in Macros

Macro statements can do arithmetic just like a calculator.

Supported operations (Fanuc style):

- Addition: +
- Subtraction: -
- Multiplication: *
- Division: /
- Absolute value: ABS[]
- Rounding: FIX[] (round down), FUP[] (round up), ROUND[]
- Square root: SQRT[]
- Trigonometry: SIN[], COS[], TAN[], ASIN[], ACOS[], ATAN[]

IMPORTANT: Angles in Fanuc macros are in degrees, not radians.

Example:

#10 = 30.0

#11 = SIN[#10]   (sin 30° = 0.5, so #11 = 0.5)

#12 = COS[#10]   (cos 30° = 0.866...)

#13 = TAN[#10]

#20 = SQRT[25.0]   (#20 = 5.0)

#21 = [10.0 + 2.0] / 3.0

10.4 Part Family concept in Macro

10.4.1 Part Family1- Washer

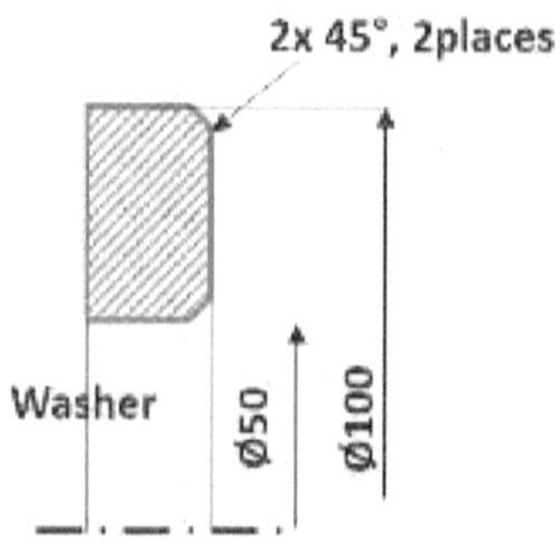

<u>Normal programming</u>

G0 X102.0 Z0.0

G01 X48.0 F0.1

G0 X94 Z1.0

G1 X100.0 Z-2.0

Z-7.0

G0 X102.0Z-5.0

G28U0W0

Using Variables

```
#100=100.0        (OD)

#101=2.0          (CHAMFER LENGTH)

#102=50.0         (ID)

#103=0.1          (FEED RATE MM/REV)

G0 X[#100+2.0] Z0.0

G01 X[#102-2.0] F#103

G0 X[#100-[2*#101+2] Z1.0

G1 X#100 Z-[#101]

Z-7.0

G0 X[#100+2.0] Z-5.0

G28U0W0
```

This program can be used for any washer in the same part family

10.4.2 Part Family2- Block

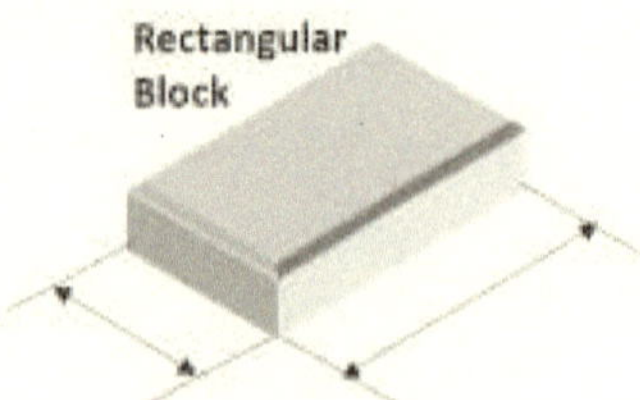

Normal programming

```
G01 X0.0 Y0.0
```

```
G01 Y50.0 F500

X100.0

Y0.0

X0.0

G01 Z5.0

G28G91Z0.
```

<u>Using Variables</u>

```
#100=100.0        (LENGTH)

#101=50.0         (WIDTH)

#103=500          (FEED RATE MM/MIN)

G01 X0.0 Y0.0

G01 Y#101 F#103

X#100

Y0.0

X0.0

G01 Z5.0

G28G91Z0.
```

This program can be used for any block in the same part family

10.5 Coordinate Calculations with Trigonometry

One of the most useful things in macro programming is automatically calculating coordinates from an angle and radius. This is critical for drilling bolt circles, machining impellers, ports arranged around a face, etc.

10.5.1 Example: Bolt Circle Pattern

Say we want N holes equally spaced on a bolt circle of radius R, centered at Xc, Yc.
For hole number i (i = 0,1,2,...N-1):

- Angle for hole i in degrees:
- θ = i * (360 / N)
- Coordinates:
- X = Xc + R * COS[θ]
- Y = Yc + R * SIN[θ]

Macro example to drill a 6-hole PCD of Ø100 mm on Z-10, feed drilling cycle:

```
( O9001 - DRILL BOLT CIRCLE )

( #1 = CENTER X, #2 = CENTER Y )

( #3 = RADIUS,   #4 = NUMBER OF HOLES )

( #5 = Z DEPTH,  #6 = SAFE Z )
```

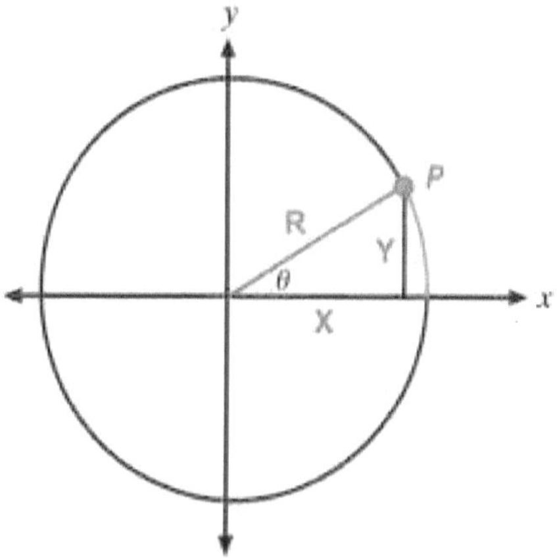

X= R Cosθ

Y= R Sinθ

```
O9001 (BOLT CIRCLE DRILL)
#20 = 0              ; (i = 0 start)
WHILE [#20 LT #4] DO1 ; (loop from i=0 until i = #4-1)
(For details, refer section 10.8.1)
 #30 = #20 * [360.0/ #4]        ; (θ = i*(360/N))
 #31 = #1 + [#3 * COS[#30]]     ;  (X coord)
 #32 = #2 + [#3 * SIN[#30]]     ;  (Y coord)
 G00 Z#6
 G00 X#31 Y#32
 G81 Z#5 R#6 F200.   (simple drill cycle example)
 G80
 #20 = #20 + 1
END1
```

M99

%

10.5.2 Calling of a macro

Format:

G65 P__ L__ A__ B__

Where P is the program number to be called

L is the repetition count (1 by default)

A, B etc. data holders which are passed to the macro.

Note: 8xxx & 9xxx series program numbers are reserved for macro programming and are having different protection level. Once protected, you cannot view or edit those programs. Please refer to the respective machine manual to find out.

Calling it from main:

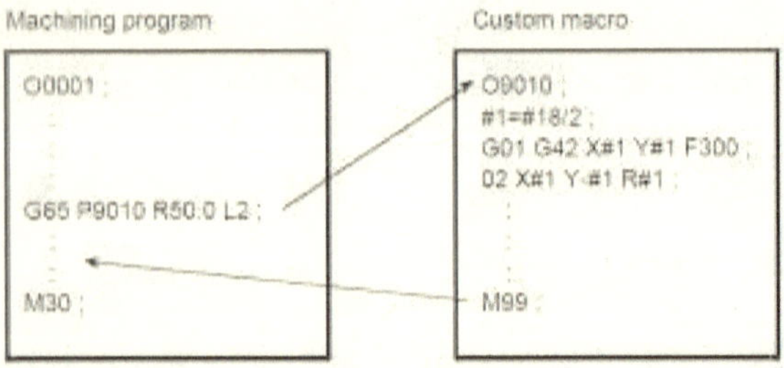

Along with G65 Alphabets other than G, L, O, N, and P can be used.

Relation between the alphabet and the variable is listed below;

Alphabet	A	B	C	D	E	F		H	I	J	K		M
Variable	#1	#2	#3	#7	#8	#9		#11	#4	#5	#6		#13

Alphabet		Q	R	S	T	U	V	W	X	Y	Z
Variable		#17	#18	#19	#20	#21	#22	#23	#24	#25	#26

Link between an alphabet and corresponding variable is defined in the controller as listed above. For example, any value written with alphabet A will be passed to #1. Same way value written with B will be passed to #2 & Z will be passed to #26

Program Example;

G65 P9010 A0.0 B0.0 C50.0 D6.0 E-10.0 F5.0

Mapping:

- A → #1 = 0.0 (center X)
- B → #2 = 0.0 (center Y)
- C → #3 = 50.0 (radius, so PCD = 100 mm)
- D → #4 = 6.0 (6 holes)
- E → #5 = -10.0 (drill depth Z)

- F → #6 = 5.0 (clearance Z)

You can see how this replaces 6 manually programmed hole positions with one smart call.

10.6 Conditional Logic

Macros are not just math calculators. You can make decisions.
This is where CNC starts feeling like a programming language.

10.6.1 IF Statements

Syntax:

```
IF [ condition ] GOTO n
```

If the condition is true , control jumps to the sequence number n.

Example:

```
#100 = 25.0

IF [#100 GT 20.0] GOTO 100

...

N100 (This block will run if #100 > 20.0)
```

Relational operators:

- EQ equal to
- NE not equal to
- GT greater than

- GE greater than or equal to
- LT less than
- LE less than or equal to

Example usage:

IF [#500 LE 0] GOTO 900 (if stored thickness ≤ 0, jump to error handler)

10.6.2 THEN form (some controls)

Some modern controls allow:

IF [#101 GT 10] THEN #101 = 0

This executes the assignment only if condition is true. Not all machines support this form; check your control.

10.7 GOTO and Labels (N numbers)

You can branch (jump) to a label:

IF [#1 LT 0] GOTO 500

...

N500

#3000 = 1 (NEGATIVE INPUT NOT ALLOWED)

That last line #3000 = 1 (message) is an *alarm you create yourself*.

- #3000 = <alarm#> (message)
  This will stop the machine with your custom

alarm text. Extremely useful for guarding against wrong inputs.

10.8 Loops

Loops let you repeat actions without copying code.

Two main patterns:

10.8.1 WHILE / END

```
#1 = 0

WHILE [#1 LT 5] DO1

 (do something)

 #1 = #1 + 1

END1
```

Notes:

- Condition is checked at the start of each loop.
- The DO1 / END1 numbers must match. You can nest loops using DO1/END1, DO2/END2, etc.

Example: Peck drilling in steps of 2 mm until final depth:

```
#100 = 0        (current depth)

#101 = 20.0     (total depth to cut)

#102 = 2.0      (peck size per pass)


WHILE [#100 LT #101] DO1
```

```
#100 = [#100 + #102]

IF [#100 GT #101] THEN #100 = #101


G01 Z[-#100] F150.

G00 Z5.0

END1
```

10.8.2 Incrementing with a Counter

This is the same idea used earlier in the bolt-circle program.

10.9 Practical Example: Parametric Slot Milling Macro

Goal: Mill a straight slot of given length, width, and depth using multiple Z passes. You call the macro with different dimensions for different parts. No need to rewrite code.

10.9.1 Macro O9100

Arguments:

- A = slot center X
- B = slot center Y
- C = slot length (along X)
- D = slot width (used maybe for tool offset check / probing)

- E = final depth (negative Z)
- F = depth per pass
- H = federate

Program:

```
O9100 (SLOT MILLING MACRO)

#1 = #1    (X center)

#2 = #2    (Y center)

#3 = #3    (slot length)

#4 = #4    (slot width - not used for path here, but could
be QC logged)

#5 = #5    (final depth, negative)

#6 = #6    (step per pass)

#11 = #11  (feed)

#20 = 0          (current depth)

#30 = [#3 / 2.0]   (half-length for +/- travel)


( rapid above slot )
```

```
G00 X#1 Y#2 Z5.0

X[#1 - #30]

WHILE [#20 GT #5] DO1    (remember #5 is negative,
e.g. -10.0)

  #20 = [#20 - #6]       (go deeper by step)

  IF [#20 LT #5] THEN #20 = #5  (limit last pass exactly at
final depth)


  G01 Z#20 F#11

  (mill along +X)

  G01 X[#1 + #30] F#11

  G00 Z5.0

X[#1 - #30]

END1

G00 Z100.0

M99

%
```

10.9.2 Calling a Macro

(T1 = 10 mm endmill)

T1 M06

S3000 M03

G90 G54

G65 P9100 A100.0 B50.0 C60.0 D12.0 E-10.0 F2.0 H300.0

Meaning:

- Slot center at X100 Y50
- Slot length = 60 mm
- Slot width = 12 mm (for info)
- Final depth Z = -10.0
- Cut in steps of 2.0 mm
- Feed = 300 mm/min

Result:

- Same macro works for any slot length, depth, or feed just by passing new values. Zero reprogramming.

10.9.3 Sample macro program to cut a circle using G01

(This program is only to understand the concepts. For circle there are qualified G-codes)

```
%
O1234(CIRCLE MACRO)
G28G91Z0
G28G91X0Y0
G0G54G90G64
T01M6
G0X0Y0
G43H01Z100.0
S1000M03
G0 Z5.0
G1 Z0 F1000
#100=100        (DIA OF CIRCLE)
#101=0          (START ANGLE)
#102=1          (ANGLE INCREMENT)
#103=#100/2     (RADIUS)
#104=360/#102 (NO. OF POINTS)
#105=0          (COUNTER SET TO 0)
#110=1000       (FEED RATE MM/MIN)
WHILE[#105LE#104]DO 1
#106=#101+#105*#102 (CURRENT ANGLE POINT)
```

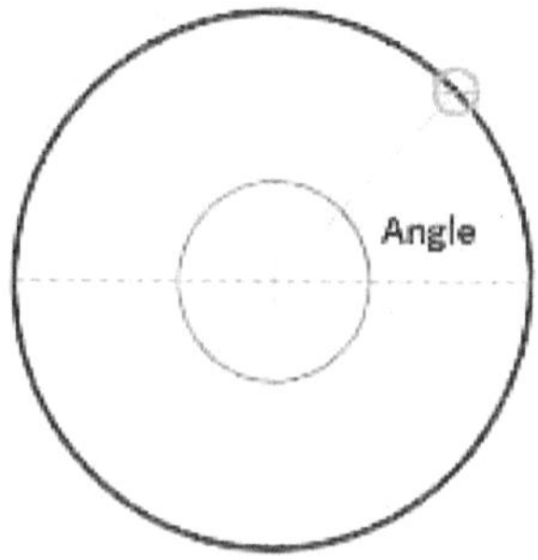

```
#107=#103*COS[#106] (CURRENT X POS)

#108=#103*SIN[#106] (CURRENT Y POS)

G64 G1 X#107 Y#108 F#110

#105=#105+1 (COUNTER INCREMENT)

END 1

G0G90X0Y0

Z100.0

G28G91Z0

G28G91X0Y0

M30

%
```

10.10 Best Practices / Safety Notes

1. **Always validate input with IF + #3000 alarm.**
   Example:
2. IF [#6 LE 0] THEN #3000 = 1 (DEPTH STEP MUST BE >0)
3. **Avoid overwriting system critical variables unless you know what you're doing.**
   Some system variables can crash the machine or mess with offsets.

4. **Document arguments at the top of every macro.**
   Treat macros like library functions. Future-you will thank you.

5. **Use non-volatile variables (#500+) for counters/probing logs that must survive power-off.**

6. **Keep angles in degrees.**
   Mixing degree-based SIN[] with CAD radian math is a classic source of mistakes. State units in comments.

PRECISION RECAP

Macro programming transforms the CNC from "execute coordinates" to "execute logic."
You get:

- Variables (local and global) to carry data
- Math and trig to calculate coordinates
- Conditional branching (IF / GOTO)
- Loops (WHILE / END)
- Modular subprogram calls (G65 Pxxxx ...)
- Custom alarms for protection

This is how you build intelligent, reusable machining cycles — and this is what separates an operator who runs parts from a manufacturing engineer who designs processes.

Chapter 11: Career Blueprint – Growing from CNC Operator to CNC Engineer

> CNC FOCUS POINTS
>
> This chapter outlines a roadmap for advancement from Entry-Level CNC Operator through to CNC Engineer, detailing roles, key skills, salary expectations, and strategies for accelerated growth.

11.1 Introduction

CNC (Computer Numerical Control) machining offers a vibrant and structured career path within modern manufacturing, blending hands-on precision with continual learning and professional growth. Unlike industries with ambiguous advancement tracks, CNC careers progress methodically, rewarding technical mastery, problem-solving, and adaptability.

The Four-Stage Career Path

- **Stage 1: CNC Operator (0–2 years)**
  Entry-level role focusing on safe machine operation and foundational quality checks.
- **Stage 2: CNC Machinist (2–5 years)**
  Skilled technician independently managing setups and optimizing processes.

- **Stage 3: CNC Programmer (5–8 years)**
  Technical expert creating and refining efficient CNC programs.
- **Stage 4: CNC Engineer (8+ years)**
  Strategic leader designing end-to-end manufacturing solutions and driving innovation.

Each stage builds critical expertise, preparing professionals for leadership, consulting, or entrepreneurial ventures in precision manufacturing.

11.2 Stage 1 — CNC Operator (Entry-Level)

Role Overview

- Salary Range: ₹2.5–₹4 lakh per annum (varies by region and sector in India)
- Time to Master: 6–18 months
- Primary Focus: Execute programmed CNC operations safely under supervision.

Core Responsibilities

- Loading/unloading workpieces with strict safety and quality adherence
- Running pre-developed CNC programs, monitoring machine performance
- Measuring parts using calipers, micrometers, height gauges

- Identifying irregularities in tool wear or machine behavior
- Documenting cycle times, tool changes, and inspection data

Essential Skills

- Blueprint reading and fundamental GD&T understanding
- Basic G- and M-code familiarity (e.g., G00, G01, M03, M05)
- Proficiency with precision metrology tools
- Routine machine maintenance (lubrication, coolant, cleaning)
- Safety protocols

Advancement Tips

- Shadow skilled colleagues to learn setups and tooling
- Study part functionality beyond drawings
- Maintain a learning journal
- Cross-train on multiple machine types
- Take online courses or certificate programs

Common Pitfalls

- Complacency in repetitive tasks

- Avoiding challenging setups
- Weak networking inside the shop
- Rushing could lead to errors and safety risks

11.3 Stage 2 — CNC Machinist

Role Overview

- Salary Range: ₹4.5–₹7 lakh per annum
- Time to Master: 2–4 years
- Focus: Independent setups, process optimization, quality assurance

Core Responsibilities

- Complete machine and tooling setups, ensure fixturing integrity
- Optimize feeds, speeds, and toolpaths at the machine
- Perform first-article inspections and troubleshoot issues
- Document setups and standard operating procedures

Key Skills

- Understanding fixtures and work holding
- Selecting and optimizing cutting tools
- Fine-tuning machining parameters

- G-code editing and Work Coordinate System (WCS) management
- Applying GD&T principles

CAM Software Exposure

- Beginners: Mastercam basics
- Advanced: SolidCAM, Esprit, Siemens NX
- Learn 2D first, then progress to 3D and multi-axis

Strategies for Growth

- Volunteer for complex projects (exotic materials, tight tolerances)
- Learn CAM programming progressively
- Build rapport with programmers and engineers
- Pursue formal diplomas or associate degrees

11.4 Stage 3 — CNC Programmer

Role Overview

- Salary Range: ₹6.0–₹12 lakh per annum
- Time to Master: 3–5 years
- Focus: Develop optimized CNC programs, simulations, and post processors

Core Responsibilities

- Generate programs from CAD models or drawings
- Plan toolpaths, verify for collisions and efficiency
- Manage machine compatibility and tooling strategies

Advanced Skills

- Multi-axis programming including 4 and 5-axis
- Parametric and logic-based macro programming
- High-speed and trochoidal machining techniques

Growth Tips

- Specialize in high-demand sectors (aerospace, medical)
- Master diverse CAM systems
- Act as a mentor to machinists
- Attend trade shows and industry webinars regularly

11.5 Stage 4 – CNC Engineer

Role Overview

- Salary Range: ₹9–₹20+ lakh per annum
- Experience: 8+ years

- Focus: Strategic process design and manufacturing leadership

Key Responsibilities

- Design comprehensive machining workflows
- Select equipment and tooling aligned with objectives
- Lead cross-department teams
- Drive manufacturing automation and Industry 4.0 adoption

Advanced Competencies

- Statistical Process Control (SPC)
- CAD-based fixture and tooling design
- Integration of robotics and IoT systems

Career Path Options

- Managerial: Plant or Manufacturing Manager
- Technical: Applications Engineer or Consultant
- Entrepreneurial: Machine shop owner or consultant

Personal Development

- Gain business acumen and project management skills

- Pursue bachelor/master's degrees and professional certifications

11.6 Professional Development & Certification

- Entry: Courses on CNC
- Advanced: CAD/ CAM certifications
- Education: ITI (Industrial Training Institute), Diplomas, Bachelors in Mechanical/Manufacturing Engineering

11.7 Essential Soft Skills

- Effective communication and technical writing
- Analytical problem-solving and root cause analysis
- Leadership and team collaboration
- Adaptability to evolving technology
- Commitment to continuous improvement

11.8 Industry Trends & Future Opportunities

Emerging*

- Hybrid additive-subtractive manufacturing
- AI-driven toolpath and process optimization

- Connected machines (IIoT) for predictive maintenance
- Advanced materials machining (composites, ceramics)

Growing Sectors

- Electric Vehicles (EVs)
- Renewable energy components
- Medical implants and devices
- Aerospace and defense

11.9 Compensation & Career Timeline

Stage	Role	Salary Range (₹) p.a	Typical Timeframe
Entry	CNC Operator	2.5–4 lakh	0–2 years
Skilled	CNC Machinist	4.5–7 lakh	2–5 years

Stage	**Role**	**Salary Range (₹) p.a**	**Typical Timeframe**
Technical	CNC Programmer	6–12 lakh	5–8 years
Professional	CNC Engineer	9–20+ lakh	8+ years
Leadership	Engineering Manager	15–30+ lakh	12+ years

Disclaimer: Salaries indicated here are only for reference and comparison. Actual industry rates may vary.

11.10 Building Your Career Strategy

Self-Assessment

- Preferred work environment and lifestyle balance
- Interests in tech, programming, automation, or operations
- Long-term financial and professional goals
- Willingness to upskill and relocate

Development Plan

- Years 1–2: Master fundamentals, safety, and entry certifications
- Years 3–5: Advance machining, basic CAM, and documentation skills
- Years 6–10: Transition to programming, project involvement, and leadership
- 10+ years: Focus on engineering, management, consulting, or entrepreneurship

Investment in Growth

- Continuous learning budget (3–5% of income)
- Courses, certifications, networking, conferences

PRECISION RECAP

The CNC career path from operator to engineer is clear, systematic, and rewarding. With commitment to technical mastery, continuous education, and strategic planning, professionals can build lasting, lucrative careers. Emerging tech and industry 4.0 integration offer exciting new opportunities for those ready to embrace change and innovation.

The future belongs to those who learn, adapt, and lead in the precision manufacturing landscape.

12. Appendix

12.1 FANUC MILL G-Code Reference

Reference- Fanuc Manual B-64694EN-2_01

G Code	Function
G00	Positioning (rapid traverse)
G01	Linear interpolation (cutting feed)
G02	Circular interpolation CW or helical interpolation CW
G03	Circular interpolation CCW or helical interpolation CCW
G04	Dwell
G04.1	G code preventing buffering
G05	AI contour control (high-precision contour control compatible command)
G05.1	AI contour control
G05.4	HRV3 on/off
G07.1	Cylindrical interpolation
G08	AI contour control (advanced preview control compatible command)
G09	Exact stop
G10	Programmable data input
G11	Programmable data input mode cancel
G16	Polar coordinates command
G15	Polar coordinates command cancel
G17	XpYp plane selection
G18	XpZp plane selection
G19	YpZp plane selection
G20 (G70)	Input in inch

G21 (G71)	Input in mm
G22	Stored stroke check function on
G23	Stored stroke check function off
G25	Spindle speed fluctuation detection on
G26	Spindle speed fluctuation detection off
G27	Reference position return check
G28	Automatic return to reference position
G28.2	In-position check disable reference position return
G29	Movement from reference position
G30	2nd, 3rd, and 4th reference position return
G30.2	In-position check disable 2nd, 3rd, or 4th reference position return
G31	Skip function
G31.8	EGB-axis skip
G33	Threading
G37	Automatic tool length measurement
G40	Tool radius/tool nose radius compensation: cancel
G41	Tool radius/tool nose radius compensation: left
G42	Tool radius/tool nose radius compensation: right
G41.1	Normal direction control on: left
G42.1	Normal direction control on: right
G40.1	Normal direction control cancel mode
G43	Tool length compensation +
G44	Tool length compensation -
G45	Tool offset: increase
G46	Tool offset: decrease
G47	Tool offset: double increase
G48	Tool offset: double decrease

G49 (G49.1)	Tool length compensation cancel
G50	Scaling cancel
G51	Scaling
G50.1	Programmable mirror image cancel
G51.1	Programmable mirror image
G50.4	Cancel synchronous control
G51.4	Start synchronous control
G50.6	Cancel superimposed control
G51.6	Start superimposed control
G50.5	Cancel composite control
G51.5	Start composite control
G52	Local coordinate system setting
G53	Machine coordinate system setting
G53.1	Tool axis direction control
G53.2	Selecting a machine coordinate system with feed rate
G53.6 (G54.1)	Tool center point retention type tool axis direction control
G54	Workpiece coordinate system 1 selection
G55	Workpiece coordinate system 2 selection
G56	Workpiece coordinate system 3 selection
G57	Workpiece coordinate system 4 selection
G58	Workpiece coordinate system 5 selection
G59	Workpiece coordinate system 6 selection
G60	Single direction positioning
G61	Exact stop mode
G62	Automatic corner override
G63	Tapping mode
G64	Cutting mode
G65	Macro call
G66	Macro modal call A
G66.1	Macro modal call B

G67	Macro modal call A/B cancel
G68	Coordinate system rotation start or 3-dimensional coordinate conversion mode on
G69	Coordinate system rotation cancel or 3-dimensional coordinate conversion mode off
G68.2	Tilted working plane indexing
G68.3	Tilted working plane indexing by tool axis direction
G68.4	Tilted working plane indexing (incremental multi-command)
G72.1	Figure copying (rotary copy)
G72.2	Figure copying (linear copy)
G73	Peck drilling cycle
G74	Left-hand tapping cycle
G75	Plunge grinding cycle
G76	Fine boring cycle
G77	Plunge direct sizing/grinding cycle
G78	Continuous-feed surface grinding cycle
G79	Intermittent-feed surface grinding cycle
G80	Canned cycle cancel
G80.4	Electronic gear box: synchronization cancellation
G81.4	Electronic gear box 1P: synchronization start
G80.5	Electronic gear box 2 pair: synchronization cancellation
G81.5	Electronic gear box 2 pair: synchronization start
G81	Drilling cycle or spot boring cycle
G81.1	High precision oscillation
G82	Drilling cycle or counter boring cycle

G83	Peck drilling cycle
G84	Tapping cycle
G84.2	Rigid tapping cycle (FS15 format)
G84.3	Left-handed rigid tapping cycle (FS15 format)
G85	Boring cycle
G86	Boring cycle
G87	Back boring cycle
G88	Boring cycle
G89	Boring cycle
G90	Absolute programming
G91	Incremental programming
G91.1	Checking the maximum incremental amount specified
G92	Setting for workpiece coordinate system or clamp at maximum spindle speed
G92.1	Workpiece coordinate system preset
G93	Inverse time feed
G94	Feed per minute
G95	Feed per revolution
G96	Constant surface speed control
G97	Constant surface speed control cancel
G96.1	Spindle indexing execution (waiting for completion)
G96.2	Spindle indexing execution (not waiting for completion)
G96.3	Spindle indexing completion check
G96.4	SV speed control mode ON
G98	Canned cycle: return to initial level
G99	Canned cycle: return to R point level
G107	Cylindrical interpolation
G160	In-feed control cancel
G161	In-feed control

2.2 FANUC LATHE G-Code Reference

Reference- Fanuc Manual B-64694EN-1_01

G code system A	G code system B	G code system C	Function
G00	G00	G00	Positioning (Rapid traverse)
G01	G01	G01	Linear interpolation (Cutting feed)
G02	G02	G02	Circular interpolation CW or helical interpolation CW
G03	G03	G03	Circular interpolation CCW or helical interpolation CCW
G04	G04	G04	Dwell
G04.1	G04.1	G04.1	G code preventing buffering
G05	G05	G05	AI contour control (command compatible with high precision contour control)
G05.1	G05.1	G05.1	AI contour control
G05.4	G05.4	G05.4	HRV3 on/off
G07.1 (G107)	G07.1 (G107)	G07.1 (G107)	Cylindrical interpolation
G08	G08	G08	AI contour control (advanced preview control compatible command)
G09	G09	G09	Exact stop

G code system A	G code system B	G code system C	Function
G10	G10	G10	Programmable data input
G11	G11	G11	Programmable data input cancel
G12.1 (G112)	G12.1 (G112)	G12.1 (G112)	Polar coordinate interpolation mode
G13.1 (G113)	G13.1 (G113)	G13.1 (G113)	Polar coordinate interpolation cancel
G17	G17	G17	XY plane selection
G18	G18	G18	XZ plane selection
G19	G19	G19	YZ plane selection
G20	G20	G70	Input in inch
G21	G21	G71	Input in mm
G22	G22	G22	Stored stroke check on
G23	G23	G23	Stored stroke check off
G25	G25	G25	Spindle speed fluctuation detection off
G26	G26	G26	Spindle speed fluctuation detection on
G27	G27	G27	Reference position check
G28	G28	G28	Return to reference position
G29	G29	G29	Movement from reference position
G30	G30	G30	2nd, 3rd, 4th reference return
G31	G31	G31	Skip function

G code system A	G code system B	G code system C	Function
G32	G33	G33	Threading
G34	G34	G34	Variable lead threading
G35	G35	G35	Circular threading CW
G36	G36	G36	Circular threading CCW
G37	G37	G37	Automatic tool offset (Z axis)
G38	G38	G38	Tool radius/tool nose radius compensation: with vector held
G39	G39	G39	Tool radius/tool nose radius compensation: corner rounding interpolation
G40	G40	G40	Tool radius/tool nose radius compensation: cancel
G41	G41	G41	Tool radius/tool nose radius compensation: left
G42	G42	G42	Tool radius/tool nose radius compensation: right
G40.1	G40.1	G40.1	Normal direction control cancel mode
G41.1	G41.1	G41.1	Normal direction control left on
G42.1	G42.1	G42.1	Normal direction control right on

G code system A	G code system B	G code system C	Function
G43.7	G43.7	G43.7	Tool offset (Bit 3 (TCT) of parameter No. 5040 must be *1.)
G50	G92	G92	Coordinate system setting or max spindle speed clamp
G50.3	G92.1	G92.1	Workpiece coordinate system preset
G50.1	G50.1	G50.1	Programmable mirror image cancel
G51.1	G51.1	G51.1	Programmable mirror image
G50.2 (G250)	G50.2 (G250)	G50.2 (G250)	Polygon turning cancel
G51.2 (G251)	G51.2 (G251)	G51.2 (G251)	Polygon turning
G50.4	G50.4	G50.4	Cancel synchronous control
G50.5	G50.5	G50.5	Cancel composite control
G50.6	G50.6	G50.6	Cancel superimposed control
G51.4	G51.4	G51.4	Start synchronous control
G51.5	G51.5	G51.5	Start composite control
G51.6	G51.6	G51.6	Start superimposed control
G52	G52	G52	Local coordinate system setting

G code system A	G code system B	G code system C	Function
G53	G53	G53	Machine coordinate system setting
G53.1	G53.1	G53.1	Tool axis direction control
G53.2	G53.2	G53.2	Selecting machine coordinate system with feed rate
G53.6	G53.6	G53.6	Tool center point retention type tool axis direction control
G54	G54	G54	Workpiece coordinate system 1 selection
G55	G55	G55	Workpiece coordinate system 2 selection
G56	G56	G56	Workpiece coordinate system 3 selection
G57	G57	G57	Workpiece coordinate system 4 selection
G58	G58	G58	Workpiece coordinate system 5 selection
G59	G59	G59	Workpiece coordinate system 6 selection
G61	G61	G61	Exact stop mode
G63	G63	G63	Tapping mode
G64	G64	G64	Cutting mode
G65	G65	G65	Macro call
G66	G66	G66	Macro modal call A
G66.1	G66.1	G66.1	Macro modal call B
G67	G67	G67	Macro modal call A/B cancel

G code system A	G code system B	G code system C	Function
G68	G68	G68	Mirror image on for double turret or balance cutting mode
G68.1	G68.1	G68.1	Coordinate system rotation start or 3D coordinate system conversion mode on
G68.2	G68.2	G68.2	Tilted working plane indexing command by tool axis direction
G68.3	G68.3	G68.3	Tilted working plane indexing command
G68.4	G68.4	G68.4	Tilted working plane indexing command (incremental multi-command)
G69	G69	G69	Mirror image off for double turret or balance cutting mode cancel
G69.1	G69.1	G69.1	Coordinate system rotation cancel or 3-dimensional coordinate system conversion mode off
G70	G70	G70	Finishing cycle
G71	G71	G71	Stock removal in turning
G72	G72	G72	Stock removal in facing
G73	G73	G73	Pattern repeating cycle

G code system A	G code system B	G code system C	Function
G74	G74	G74	End face peck drilling cycle
G75	G75	G75	Outer diameter/inner diameter drilling cycle
G76	G76	G76	Multiple-thread cutting cycle
G80	G80	G80	Canned cycle cancel for drilling Electronic gear box: synchronization cancellation
G81.1	G81.1	G81.1	High precision oscillation function
G80.4	G80.4	G80.4	Electronic gear box: synchronization cancellation
G80.5	G80.5	G80.5	Electronic gear box 2 pair: synchronization start
G81.5	G81.5	G81.5	Electronic gear box 2 pair: synchronization start
G81	G81	G81	Spot drilling (FS15・T format)
G82	G82	G82	Counter boring (FS15・T format)
G83	G83	G83	Cycle for face drilling
G83.1	G83.1	G83.1	High-speed peck drilling cycle (FS15・T format)

G code system A	G code system B	G code system C	Function
G83.5	G83.5	G83.5	High-speed peck drilling cycle
G84	G84	G84	Cycle for face tapping
G84.2	G84.2	G84.2	Rigid tapping cycle (FS15・T format)
G85	G85	G85	Cycle for face boring
G87	G87	G87	Cycle for side drilling
G87.5	G87.5	G87.5	High-speed peck drilling cycle
G87.6	G87.6	G87.6	Peck drilling cycle
G88	G88	G88	Cycle for side tapping
G89	G89	G89	Cycle for side boring
G90	G77	G20	Outer diameter/inner diameter cutting cycle
G92	G78	G21	Threading cycle
G94	G79	G24	End face turning cycle
G91.1	G91.1	G91.1	Maximum specified incremental amount check
G96	G96	G96	Constant surface speed control
G97	G97	G97	Constant surface speed control cancel
G96.1	G96.1	G96.1	Spindle indexing execution (waiting for completion)
G96.2	G96.2	G96.2	Spindle indexing execution (not waiting for completion)

G96.3	G96.3	G96.3	Spindle indexing completion check
G96.4	G96.4	G96.4	SV speed control mode ON
G98	G94	G94	Feed per minute
G99	G95	G95	Feed per revolution
-	G90	G90	Absolute programming
-	G91	G91	Incremental programming
-	G98	G98	Canned cycle: return to initial level
-	G99	G99	Canned cycle: return to R point level

12.3 List of program examples

M/C	Category	Detail	codes	Section
All	Dwell	Dwell time	G04	2.11
All	Unit Selection	Inch / Metric	G20/ G21	2.13.2
Mill	Subprogram	Face milling	M98/ M99	2.15.3
Mill	Subprogram	Drilling pattern	M98/ M99	2.15.4
Mill	Subprogram	Hole interpolation	M98/ M99	2.15.5
Mill	Subprogram	Nesting	M98/ M99	2.15.6
Mill	CRC	Manual Compensation	G40	2.16.8
Mill	CRC	Compensation Left /Right	G41/ G42	2.16.9
Mill	CRC	Simplified	G41/ G42	2.16.10
Mill	CRC	Circular boss	G41	2.16.11
Mill	Simplify	Auto radius	R	2.17.3
Mill	Simplify	Auto Chamfer	C	2.17.4
Mill	Cycles	Fine boring	G76	2.18.5.6
Mill	Cycles	Drilling	G83	2.18.6.1
Mill	Cycles	Tapping	G84	2.18.6.2
Turn	Linear	Facing	G01/ G96	2.26.5
Turn	Simplify	Direct angle	A	2.28.8
Turn	Cycles	Stock removal-Turn	G71	2.29.2
Turn	Cycles	Stock removal-Face	G72	2.29.3

M/C	Category	Detail	codes	Section
Turn	Cycles	End Face Drilling	G74	2.29.5
Turn	Cycles	End Face Grooving	G74	2.29.5
Turn	Cycles	OD/ID Grooving	G75	2.29.5
Turn	Cycles	Threading	G76	2.29.7
Turn	Cycles	Turning rough /finish	G71/ G70	2.29.8
Turn Mill	Mill	Face milling	G12.1/ G112	2.30.8
Turn Mill	Mill	Axial drilling	M154/ M155	2.30.9
Turn Mill	Mill	Radial drilling	M154/ M155	2.30.10
5X Mill	Multi face machining	Tilted plane	G68.2	2.32.7.3
Turn	Macro	Part family / Washer	#	10.4.1
Mill	Macro	Part family / Block	#	10.4.2
Mill	Macro	Bolt circle	#	10.5.1
Mill	Macro	Slot milling	#	10.9.1
Mill	Macro	Slot milling	G65	10.9.2
Mill	Macro	Circle using G01	G65	10.9.3
Mill	Macro	Sphere milling- Z level	Siemen s	12.7.1
Mill	Macro	Sphere milling - Radial	Siemen s	12.7.2

12.4 A to Z of CNC: The Complete Alphabet That Speaks Precision

Master the Language of Modern Manufacturing

CNC programming is more than a series of codes—it's a sophisticated language that transforms digital designs into tangible, precision-engineered components. Like any language, it has its own alphabet, grammar, and syntax, where each letter carries precise meaning, dictating how machines cut, move, and create with tolerances as fine as thousandths of an inch. Whether you're an aspiring programmer, seasoned machinist, or engineering student, mastering this CNC alphabet is your foundation for success in modern manufacturing. Let's decode the A to Z of CNC, exploring each letter's significance and its real-world applications.

A – A-Axis: The First Rotary Revolution

Definition: Rotary motion around the X-axis, enabling tilt for multi-axis machining.

The A-axis introduces true 4-axis and 5-axis machining, allowing parts to tilt and rotate without manual repositioning. This rotational capability around the X-axis enables complex geometries in a single setup.

Real-World Applications:

- **Aerospace**: Turbine blades with intricate angular cuts.

- **Medical**: Hip implants with curved surfaces.
- **Automotive**: Intake manifolds with angled ports.

Programming Example:

G00 A45.0; Rotate A-axis to 45 degrees

G01 Z-0.500 F10.0; Machine angled surface

Pro Tip: Account for tool length compensation, as the effective tool length changes with A-axis rotation. Use simulation software to verify toolpaths and avoid collisions.

B – B-Axis: Y-Axis Rotation Mastery

Definition: Rotary motion around the Y-axis, critical for advanced machining centers and turn-mill machines.

The B-axis enables rotation around the Y-axis, supporting complex angular operations and five-sided machining, especially in turn-mill setups where milling and turning occur simultaneously.

Applications:

- **Complex Parts**: Engine blocks with angled cylinders.
- **Pump Housings**: Multiple inlet/outlet ports at various angles.
- **Molds**: Components requiring draft angles and intricate geometries.

Programming Considerations:

- Coordinate system rotation affects subsequent moves.
- Tool collision detection is critical due to dynamic positioning.
- Ensure work holding accommodates the full rotational range.

C – C-Axis: The Turning Revolution

Definition: Rotary motion around the Z-axis, transforming lathes into versatile turn-mill machines.

The C-axis enables rotational positioning on lathes, allowing milling operations on turned parts without removing them from the chuck, revolutionizing turning capabilities.

Game-Changing Capabilities:

- **Cross-Drilling**: Radial holes in shafts.
- **Keyway Milling**: Precise slots in rotating components.
- **Polygon Turning**: Square or hexagonal shapes on round stock.
- **Helical Operations**: Threads, gears, and spiral features.

Programming Example:

```
G00 C90.0    ; Index C-axis to 90 degrees
M19          ; Orient spindle for milling
G01 X1.0 F5.0    ; Mill across diameter
```

D – Diameter Offset: Precision in Turning

Definition: Compensation values for tool radius or wear in turning operations.

D-offsets store tool radius values in turning, ensuring accurate diameter control. Unlike milling, where radius compensation is common, turning focuses on diameter precision.

Critical Applications:

- **Precision Shafts**: Bearing journals with ±0.0002" tolerances.
- **Hydraulic Cylinders**: Piston bores with specific surface finishes.
- **Medical Components**: IV connector threads with critical dimensions.

Offset Management:

```
T0101    ; Tool 1, Offset 1
G42      ; Diameter compensation right
G01 X2.000 Z-1.500 F0.008    ; Turn to 2.000" diameter
```

Best Practice: Regularly verify D-offsets using touch-off procedures or automated measurement systems to maintain precision.

E – Extended Parameters: The Flexible Variable

Definition: Multi-purpose parameter for specialized cycles and extended functionality.

The E-parameter is a versatile tool in modern CNC controllers, used for various purposes depending on the machine and operation.

Common Applications:

- Thread pitch in threading cycles.
- Dwell time extensions for specialized operations.
- Feed-per-tooth values in advanced milling cycles.
- Secondary parameters in custom macro programs.

Threading Example:

G76 P010060 Q100 R0.05 E0.05 ; Threading cycle with 0.05" finishing allowance

Controller Variations:

- **Fanuc**: Additional cycle parameters.
- **Haas**: Secondary feed rates or timing values.
- **Siemens**: Extended interpolation parameters.

- **Mazak**: Custom cycle modifiers.

F – Feed Rate: The Rhythm of Machining

Definition: Controls the rate at which cutting tools advance through material, measured in inches per minute (IPM) or mm/min.

Feed rate is the heartbeat of CNC machining, influencing tool life, surface finish, dimensional accuracy, and cycle time. Optimizing feed rates is critical for performance.

Feed Rate Strategies:

- **Roughing**:
    - Aluminum: 200–800 IPM with flood coolant.
    - Steel: 50–300 IPM, depending on tool size.
    - Titanium: 20–100 IPM with specialized tooling.
- **Finishing**:
    - Surface finish priority: 10–50 IPM with sharp tools.
    - Dimensional accuracy: Consistent feeds to prevent deflection.
    - Tool life: Balanced speed for longevity.

Advanced Feed Control:

G01 X1.0 Y1.0 F100.0 ; Linear move at 100 IPM

G02 X2.0 Y1.0 I0.5 J0 F80.0 ; Slower feed for circular interpolation

Optimization Matrix:

- Material hardness vs. feed rate.
- Tool diameter vs. chip load.
- Surface finish requirements vs. cutting speed.

G – G-Codes: The Grammar of Motion

Definition: Modal commands defining fundamental machine motions and operations.

G-codes are the verbs of CNC programming, dictating machine movements and remaining active until changed, forming the backbone of program structure.

Essential G-Codes:

- **Positioning**:
  - G00: Rapid traverse (non-cutting).
  - G01: Linear interpolation (straight-line cutting).
  - G02/G03: Clockwise/counterclockwise circular interpolation.

- **Canned Cycles**:
  - G81: Simple drilling.
  - G83: Peck drilling for deep holes.
  - G84: Rigid tapping.
  - G85: Reaming/boring.
- **Coordinate Systems**:
  - G54–G59: Work coordinate systems.
  - G90/G91: Absolute/incremental positioning.

Programming Best Practices:

```
G00 G90 G54    ; Rapid, absolute, work coordinate 1
G01 Z-0.100 F10.0    ; Plunge at controlled feed
G41 D01        ; Cutter compensation left, offset 1
```

H – Height Offsets: Tool Length Management

Definition: Compensation values for tool length, stored in controller memory.

H-offsets manage tool length differences, enabling multi-tool operations to machine to the same Z-coordinates, critical for precision and efficiency.

Setup Process:

- Select a master tool (typically the longest).

- Measure tools using touch-off or presetters.
- Calculate and enter offsets automatically or manually.
- Verify with test cuts.

Programming Integration:

T01 M06 ; Change to tool 1

G43 H01 ; Apply height offset 1

G00 Z1.0 ; Rapid to compensated Z position

Advanced Applications:

- Dynamic tool length measurement via on-machine probing.
- Tool wear compensation with automatic adjustments.
- Broken tool detection through length verification.

I – Interpolation Parameter (X-Direction)

Definition: X-axis distance from arc start point to center in circular interpolation.

The I-parameter, alongside J and K, defines arc centers for smooth curves and complex toolpaths in circular and helical interpolation.

Arc Programming:

- **Center Point Method**:

G02 X2.0 Y2.0 I1.0 J0 F20.0 ; Arc with center 1.0" in +X from start

- **Radius Method**:

G02 X2.0 Y2.0 R1.0 F20.0 ; Same arc using radius specification

Considerations:

- Verify arc direction (G02 clockwise, G03 counterclockwise).
- Ensure start and end points match for full circles.
- Use I, J, K for helical interpolation in 3D paths.

J – Interpolation Parameter (Y-Direction)

Definition: Y-axis distance from arc start point to center in circular interpolation.

The J-parameter complements I for precise circular interpolation, critical for contour machining and smooth surface generation.

Applications:

- **Contour Machining**:
- G01 X0 Y0 ; Move to arc start

G02 X1.414 Y1.414 I1.0 J1.0 F15.0 ; Quarter circle with 1.414" radius

- **Blended Corners**: Smooth transitions for improved surface finish and tool life.

Verification:

- Arc center: $\sqrt{(I^2 + J^2)}$ = radius.
- Direction: Apply right-hand rule.
- Full circle: Validate 360° rotation.

K – Interpolation Parameter (Z-Direction)

Definition: Z-axis distance to arc center, enabling helical and 3D circular interpolation.

The K-parameter extends circular interpolation into 3D, supporting helical operations like thread milling and complex contours.

Applications:

- **Thread Milling**:

G02 X1.0 Y0 Z-0.050 I-0.5 J0 K0 F10.0 ; Helical interpolation for thread

- **Spiral Ramps**: Gradual Z-axis engagement for deep pockets.
- **3D Contouring**: Sculptured surfaces for molds and aerospace parts.

L – Repeat: The Efficiency Multiplier

Definition: Specifies repetition count for canned cycles and subroutine calls.

The L-parameter enhances efficiency by reducing repetitive code in operations like drilling or subprogram execution.

Applications:

- **Bolt Circle Drilling**:

```
G81 X1.0 Y0 Z-0.5 R0.1 L8    ; Drill 8 holes in circular pattern
```

- **Subroutine Calls**:

```
M98 P1000 L5    ; Call subprogram O1000 five times
```

Strategies:

- Use for pattern drilling, roughing passes, or quality operations.
- Combine with variables for parametric programming.

M – M-Codes: The Machine's Vocabulary

Definition: Miscellaneous functions controlling machine peripherals and operations.

M-codes manage auxiliary functions like spindles, coolant, and program flow, enabling seamless machining operations.

Essential M-Codes:

- **Spindle:**
  - M03: Clockwise rotation.
  - M04: Counterclockwise rotation.
  - M05: Spindle stop.
- **Coolant:**
  - M07: Mist coolant.
  - M08: Flood coolant.
  - M09: Coolant off.
- **Tool Management:**
  - M06: Tool change.
  - M19: Spindle orient.
- **Program Control:**
  - M30: Program end/reset.
  - M01: Optional stop.
  - M00: Mandatory stop.

Advanced Example:

```
M98 P2000   ; Call subprogram O2000
M99         ; Return from subprogram
```

N – Block Number: Program Organization

Definition: Sequential line numbers for organizing and identifying program blocks.

Block numbers enhance readability, debugging, and program modification.

Numbering Strategies:

- **Incremental**:
- N10 G00 G90 G54
- N20 T01 M06

N30 G43 H01

- **Functional Grouping**:
    - N100–199: Setup.
    - N200–299: Roughing.
    - N300–399: Finishing.

Best Practices:

- Use consistent increments (e.g., 5 or 10).
- Group related operations.
- Leave gaps for future edits.

O – Program Number: Digital Identity

Definition: Unique identifier for main programs and subprograms.

O-numbers organize programs and subroutines, enabling efficient program management.

Organization Systems:

- **Job-Based**:
    - O1000–1099: Job #1.
    - O9000–9999: Standard subroutines.
- **Part-Based**:
- O0001 (MAIN PROGRAM - HOUSING)
- M98 P1001   ; Call roughing subroutine
- M30
- O1001 (ROUGHING SUBROUTINE)

M99

Version Control:

- O1001: Original.
- O1002: Revision 1.

P – Subprogram Call: Modular Programming

Definition: Parameter for specifying subprogram numbers in call commands.

P enables modular programming, reducing code length and improving maintainability.

Applications:

- **Parametric Machining**:
- #1 = 0.500   ; Hole diameter

M98 P2000   ; Call drilling subroutine

- **Pattern Generation**: Bolt circles, linear arrays, or spirals.

Q – Parameter in Canned Cycles: Precision Control

Definition: Incremental values for drilling and machining cycles.

Q provides fine control in canned cycles, especially for deep hole drilling.

Peck Drilling Example:

G83 X1.0 Y1.0 Z-2.0 R0.100 Q0.200 F10.0   ; Peck 0.200" increments

Benefits:

- Chip evacuation in deep holes.
- Reduced tool wear and heat buildup.
- Improved hole quality and accuracy.

R – Radius and Retract: Dual-Purpose Parameter

Definition: Defines arc radius or retract height in canned cycles.

R is versatile, used in circular interpolation or drilling cycles.

Examples:

- **Circular Interpolation**:

G02 X2.0 Y0 R1.0 F15.0 ; Arc with 1.0" radius

- **Canned Cycle Retract**:

G81 X1.0 Y1.0 Z-0.500 R0.100 F8.0 ; Retract to Z0.100

Strategies:

- Ensure safe clearance heights.
- Verify arc direction for multi-solution cases.

S – Spindle Speed: The Power Behind Precision

Definition: Controls spindle rotation in revolutions per minute (RPM).

Spindle speed affects material removal, tool life, and surface finish.

Speed Optimization:

- **Aluminum**:
- S2400 M03 ; 2400 RPM for 1/2" end mill

G01 Z-0.250 F48.0

  - Surface speed: 400–800 SFM.
  - Coolant: Flood or mist.

- **Stainless Steel**: 200–400 SFM, flood coolant.
- **Titanium**: 100–250 SFM, carbide tools with coatings.

Advanced Control:

```
G96 S200 M03    ; Constant surface speed
G97 S1500 M03   ; Constant RPM
```

T – Tool Call: The Workhorse Selection

Definition: Specifies the tool number for spindle loading.

T commands enable multi-tool operations with automatic tool changes.

Example:

```
T05 M06    ; Load tool #5
G43 H05    ; Apply height offset
S1800 M03  ; Start spindle
```

Tool Library:

- T01–T10: Roughing tools.
- T21–T30: Finishing tools.
- T31–T40: Drills/taps.

Best Practices:

- Standardize T and H numbers.

- Verify tool changes with simulation.

U – Incremental X: Relative Motion Control

Definition: Incremental X-axis movement, common in turning.

U simplifies turning by specifying relative movements.

Example:

G00 X2.0 Z1.0 ; Start position

G01 U-0.100 F0.008 ; Move 0.100" closer to centerline

Applications:

- Incremental turning for depth control.
- Threading with consistent engagement.

V – Incremental Y: Precision Positioning

Definition: Incremental Y-axis movement for milling operations.

V enables relative positioning for milling patterns.

Example:

G00 X1.0 Y1.0 ; Slot start

G01 V2.0 F20.0 ; Cut slot 2.0" in +Y

Applications:

- Grid drilling.
- Pocket clearing.

- Surface texturing.

W – Incremental Z: Depth Control Mastery

Definition: Incremental Z-axis movement for turning and drilling.

W is critical for depth control in turning operations.

Example:

```
G00 X3.0 Z0.100    ; Position for facing
G01 W-0.050 F0.005 ; Advance 0.050" deeper
```

Applications:

- Facing operations.
- Threading depth control.
- Deep hole drilling.

X – X-Axis: The Foundation of Motion

Definition: Primary linear axis for left-to-right movement (milling) or diameter control (turning).

Milling:

- X: Left-to-right.
- Resolution: 0.0001" or 0.001mm.

Turning:

- X: Radial direction.
- X2.0 = 2.0" diameter (diameter programming).

Example:

G00 X1.2345 Y2.3456   ; Precise positioning

Applications:

- Mirror imaging.
- Scaling operations.

Y – Y-Axis: Perpendicular Precision

Definition: Linear axis for front-to-back movement in milling.

Y enables 2D/3D machining with X and Z.

Example:

G01 X1.0 Y1.0 F15.0   ; Linear move

G02 X2.0 Y2.0 I0.5 J0.5 F15.0   ; Circular arc

Strategies:

- Optimize for climb milling.
- Ensure fixture clearance.

Z – Z-Axis: The Dimension of Depth

Definition: Linear axis controlling vertical movement and depth of cut.

Z is critical for depth control, with errors potentially causing crashes.

Example:

G00 Z1.0 ; Safe clearance

G01 Z-0.250 F10.0 ; Controlled plunge

Strategies:

- Positive Z for clearance, negative for cutting.
- Use consistent Z-zero references.

Advanced Programming Concepts

Modal vs. Non-Modal Commands

- **Modal** (stay active): G00, G01, G41, G43.
- **Non-Modal** (execute once): G04 (dwell), G28 (reference return).

Parameter Programming

Use variables for flexibility:

#1 = 1.000 ; Hole diameter

G81 X[#1] Y[#1] Z-0.500 R0.100

Coordinate System Management

- G54–G59: Work offsets for multiple setups.

G54 ; Select work coordinate 1

G00 X0 Y0 Z1.0 ; Position at part zero

Industry-Specific Applications

- **Aerospace**: 5-axis machining, titanium processing, ±0.0005" tolerances.
- **Medical**: Micro-machining, biocompatible materials, strict validation.
- **Automotive**: High-volume production, robotic automation.
- **Mold and Die**: Hardened steels, mirror finishes.

Programming Efficiency Tips

Code Organization

- **Header**:
- O0001 (PART: HOUSING-001 REV-C)

(PROGRAMMER: J.SMITH DATE: 03/15/2024)

- **Tool Information**:
- (T01 - 1/2" END MILL - ROUGH)

(T02 - 1/4" END MILL - FINISH)

Error Prevention

- Verify tools, offsets, and coordinates.
- Use simulation software to detect collisions.

Common Errors

- **Radius Too Small**:

- ; ❌ Wrong
- G02 X1.5 Y1.5 I0.1 J0.1
- ; ✅ Correct

  G02 X1.5 Y1.5 I0.5 J0.5

Troubleshooting Guide

- **Radius Too Small**: Adjust I, J, K or use R-method.
- **Axis Travel Limit**: Verify part position and offsets.
- **Tool Offset Error**: Match T and H numbers.

Optimization:

- Minimize air cutting.
- Reduce tool changes.
- Optimize feed rates and toolpaths.

Future of CNC Programming

Industry 4.0

- **Smart Manufacturing**: IoT for real-time monitoring, AI for predictive maintenance.
- **Digital Twins**: Virtual machine simulations.
- **Adaptive Control**: Automatic parameter optimization.

Emerging Technologies

- **Hybrid Manufacturing**: Additive + subtractive processes.
- **Advanced Controls**: Touch screens, augmented reality, voice commands.

Learning Resources and Career Development

Key Takeaways: Mastering the CNC Alphabet

Foundation Principles:

- **Precision**: Every command impacts the final product.
- **Safety**: Proper motion commands prevent accidents.
- **Efficiency**: Organized programs reduce errors and cycle times.
- **Learning**: Stay current with evolving technology.

Programming Excellence Checklist:

- **Before Coding**: Understand part requirements, select tools, plan sequence.
- **During Coding**: Use consistent formatting, verify offsets, include safety procedures.
- **After Coding**: Simulate for collisions, review with peers, document lessons.

Career Success Strategies:

- Master G/M codes and advanced techniques.
- Pursue certifications and networking.
- Develop business acumen for cost and quality management.

Conclusion: From Alphabet to Artistry

The CNC alphabet—from A-axis to Z-axis—is a powerful language that turns digital designs into precision components powering industries like aerospace, medical, and automotive. Mastering this alphabet requires practice, real-world application, and continuous learning. As Industry 4.0 and technologies like AI and hybrid manufacturing evolve, these fundamentals remain the bedrock of precision manufacturing.

Whether you're a beginner or a seasoned professional, each letter in the CNC alphabet is a tool for creativity and innovation. Speak this language fluently, and you'll shape the future of manufacturing—one precise cut at a time.

12.5 Program example Milling

Part drawing

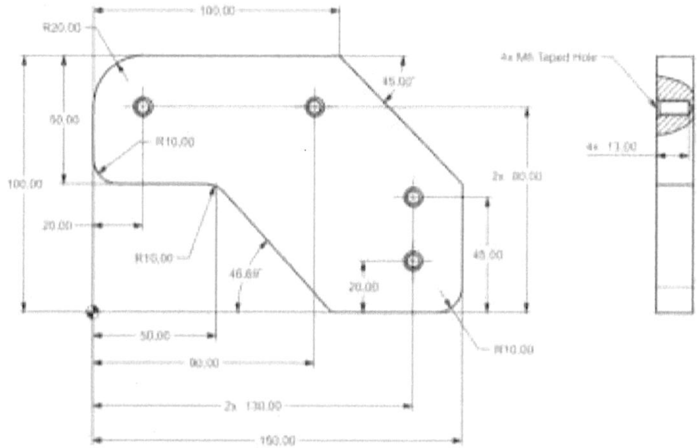

O1234 (MILLING PROGRAM)

```
G28G91Z0

G28G91X0Y0

G21 G17

G40 G54 G94 G80 G90 G0

N1T01M06 (DIA10 ENDMILL)

X50.Y110.

S5000 M03

G43H01Z100.

Z-5.M08

G01 G41 Y100. F1000
```

```
X100.

X150. Y50.

Y10.

G02 X140. Y0 R10.

G01 X97.134

X50. Y50. R10.

X10.
G02 X0 Y60. R10.

G01 Y80.

G02 X20. Y100. R20.

G01 X50.

G0 G40 Y110.

G28G91Z0

G28G91X0Y0

N2T02M06 (CENTER DRILL)

X20.Y80.

S7500 M03

G43H01Z100.M08

G81 Z-2. R2. F400

X90. Y80.
```

```
X130. Y45.

Y20.

G80

G28G91Z0

G28G91X0Y0

G28G91Z0

G28G91X0Y0

N3T03M06 (DIA5 DRILL)

X20.Y80.

S7500 M03

G43H01Z100.M08

G83 Z-13. Q5. R2. F400

X90. Y80.

X130. Y45.

Y20.

G80

G28G91Z0

G28G91X0Y0

N4T04M06 (M6 X 1P TAP)

X20.Y80.
```

```
S500 M03
G43H01Z100.M08
G84 Z-12. P3 R2. F500
X90. Y80.
X130. Y45.
Y20.
G80
M05
M09
G28G91Z0
G28G91X0Y0
M30
```

12.6 Program example Turning

Part drawing

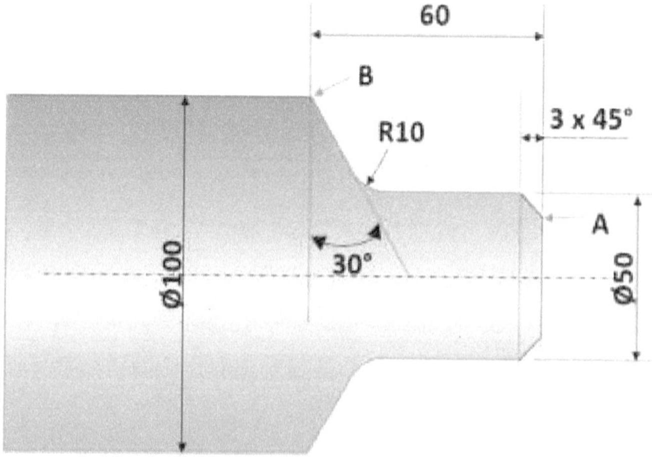


```
O1234 (TURNING PROGRAM)

G21 G18

G40 G54 G99 G80

N1 (FACE & OD ROUGH)

G28U0.W0.

G0 T0101

G97 S1000 M3

G0 G54 G40 X106. Z5. M8

G50 S2500

G96 S150
```

```
Z1.
G01 X-1. F0.3
G0 W2.
X106.
Z0.1
G01 X-1.
G0 Z2. X104.
G71 U1.0 R1.0
G71 P10 Q20 U0.4 W0.1 F0.3
N10 G00 X44. Z2.
        G01 Z0.02 F0.15
        X44. R0.1
        X50. Z-3. R0.1
        A180.0 R10.0
        X100.0 Z-60.0 A120.0
        Z-80.0
N20 X102.0
G28U0.W0.
N2 (FACE & OD FINISH)
G28U0.W0.
```

```
G0 T0202
G97 S1000 M3
G0 G54 G40 X106. Z5. M8
G50 S2500
G96 S150
Z0.
G01 X-1. F0.1
G0 W2.
X106.
G0 Z2. X104.
N10 G00 X44. Z2.
    G01 G42 Z0.02 F0.15
    X44. R0.1
    X50. Z-3. R0.1
    A180.0 R10.0
    X100.0 Z-60.0 A120.0
    Z-80.0
N20 G40 X102.0
G0 Z5.
G28U0.W0.
```

12.7 Sphere machining macro program (Siemens)

12.7.1 Method1 : Finishing with Z-level

```
WORKPIECE(,"",,"CYLINDER",64,50.1,-30,-80,100)

;SUPA G0 Z0 ;MACHINE HOME POSITION

T="CUTTER 60" M6 ;TOOL CALL

S1000 M03

R10=100 ;SPHERE DIA.

R11=60 ;ENDMILL DIA

R12=R10/2 ;SHERE RADIUS

R13=50 ;START Z, VALUE >0

R14=0 ;END Z, VALUE >=0

R20=3 ;DEPTH OF CUT, AP

R21=10 ;ENTRY RADIUS

R13=R13-R20 ; FIRST DOC

R22=R11/2 ;ENDMILL RADIUS

G54

START1:

R24=SQRT(POT(R12)-POT(R13)) ; X

R24=R24+R22 ;COMPENSATED X
```

```
G0X=R24+R21 Y=R21

Z=R13

STOPRE

G03 X=R24 Y0 CR=R21 F10000

G02 I=-R24 K=0

G03 X=R24+R21 Y=-R21 CR=R21

R13=R13-R20 ; DOC INCREMENT

IF R13>=R14 GOTOB START1

G0Z300 ; Z MAX

M00

T="BALL_END_CYL10" M6 ;TOOL CALL

S3000 M03

R10=100 ;SPHERE DIA.

R11=10 ;BALL ENDMILL DIA

R12=(R10+R11)/2 ;COMPENSATED SHERE RADIUS

R13=55 ;START Z, VALUE >0

R14=50 ;END Z, VALUE >=0

R20=0.2 ;DEPTH OF CUT, AP

R21=20 ;ENTRY RADIUS

TRANS Z=-5 ; SHIFT TOOL REFERENCE TO BALL CENTER
```

```
R13=R13-R20 ; FIRST DOC

G54

START1:

R24=SQRT(POT(R12)-POT(R13)) ; X

G0X=R24+R21 Y=R21

Z=R13

STOPRE

G03 X=R24 Y0 CR=R21 F10000

G02 I=-R24 K=0

G03 X=R24+R21 Y=-R21 CR=R21

R13=R13-R20 ; DOC INCREMENT

IF R13>=R14 GOTOB START1

G0Z300 ; Z MAX

M00

T="BALL_END_CYL10" M6 ;TOOL CALL

S3000 M03

R10=100 ;SPHERE DIA.

R11=10 ;BALL ENDMILL DIA

R12=(R10+R11)/2 ;COMPENSATED SHERE RADIUS

R13=50 ;START Z, VALUE >0
```

```
R14=0 ;END Z, VALUE >=0

R20=1 ;DEPTH OF CUT, AP

R21=10 ;ENTRY RADIUS

TRANS Z=-5 ; SHIFT TOOL REFERENCE TO BALL CENTER

R13=R13-R20 ; FIRST DOC

G54

START1:

R24=SQRT(POT(R12)-POT(R13)) ; X

G0X=R24+R21 Y=R21

Z=R13

STOPRE

G03 X=R24 Y0 CR=R21 F10000

G02 I=-R24 K=0

G03 X=R24+R21 Y=-R21 CR=R21

R13=R13-R20 ; DOC INCREMENT

IF R13>=R14 GOTOB START1

G0Z300 ; Z MAX

M30
```

12.7.2 Method2: Radial finishing

```
WORKPIECE(,"",,"CYLINDER",64,50.1,-30,-80,100)

;SUPA G0 Z0 ;MACHINE HOME POSITION

T="CUTTER 60" M6 ;TOOL CALL

S3000 M03

R10=100 ;SPHERE DIA.

R11=60 ;ENDMILL DIA

R12=R10/2 ;COMPENSATED SHERE RADIUS

R13=50 ;START Z, VALUE >0

R14=0 ;END Z, VALUE >=0

R20=2 ;DEPTH OF CUT, AP

R21=10 ;ENTRY RADIUS

R13=R13-R20 ; FIRST DOC

R22=R11/2 ;ENDMILL RADIUS

G54

START1:

R24=SQRT(POT(R12)-POT(R13)) ; X

R24=R24+R22 ;COMPENSATED X

G0X=R24+R21 Y=R21

Z=R13
```

```
STOPRE

G03 X=R24 Y0 CR=R21 F10000

G02 I=-R24 K=0

G03 X=R24+R21 Y=-R21 CR=R21

R13=R13-R20 ; DOC INCREMENT

IF R13>=R14 GOTOB START1

G0Z300 ; Z MAX

M00

M00

T="BALL_END_CYL10" M6 ;TOOL CALL

S3000 M03

R10=100 ;SPHERE DIA.

R11=10 ;BALL ENDMILL DIA

R12=(R10+R11)/2 ;COMPENSATED SHERE RADIUS

R9=10 ;START DIA

R13=SQRT(POT(R12)-POT(R9/2)) ;START Z, VALUE >0

R14=20 ;END Z, VALUE >=0

R20=3 ;ANGLE INCREMENT

R21=10 ;ENTRY RADIUS

TRANS Z=-5 ; SHIFT TOOL REFERENCE TO BALL CENTER
```

```
;R13=R13-R20 ; FIRST DOC
G54 G0 Z=R12+5
R25=0 ;START ANGLE
R29=R9/2 ;START RADIUS
WHILE R25<=360
R30=R29*COS(R25) ;X
R31=R29*SIN(R25) ;Y
R32=R13
G0 X=R30 Y=R31
G01 Z=R13 F10000 ; CUT START
WHILE R32>=R14
R36=SQRT(POT(R12)-POT(R32))
R30=R36*COS(R25) ;X
R31=R36*SIN(R25) ;Y
G64 G01 X=R30 Y=R31 Z=R32 F10000
R32=R32-0.1
ENDWHILE
G0 Z=R12+5
R25=R25+R20
ENDWHILE
M30
```

12.8 Comparison of few commands between Fanuc & Siemens

Quick Reference Table

Function	Fanuc	Siemens	Notes
Unit Selection - Metric	G21	G71 or G710	G710 includes feed rate conversion
Unit Selection - Inch	G20	G70 or G700	G700 includes feed rate conversion
Corner Rounding	,R (with comma)	RND=	Siemens has modal option: RNDM=
Chamfering	,C (with comma)	CHR= or CHF=	CHR=path length, CHF=perpendicular
Circular Arc Radius	R	CR=	Different from chamfering in Siemens
Circular Arc Center	I J K	I J K or I= J= K=	Both use IJK for center offset

Function	Fanuc	Siemens	Notes
Angle Programming	A (optional feature)	ANG=	Siemens is standard; Fanuc requires "Direct Drawing Input"
Macro Variables	#1, #100, etc.	R1, R2, etc.	Fanuc uses #, Siemens uses R
Subprogram Call	M98 P100	L100 or program name	Fanuc uses M-code, Siemens uses L address
Subprogram Return	M99	M17	Both return to calling program
Repeat Subprogram	M98 P100 L4	P command	Different syntax structures
Transformation On	G12.1 or G112	TRANSMIT or TRANSMIT(n)	For milling on turned parts
Transformation Off	G13.1 or G113	TRAFOOF	NOT "TRANSMIT OFF"
5-Axis TCP (RTCP)	G43.4 (Type I)	TRAORI	Tool Center Point Control for

Function	Fanuc	Siemens	Notes
	or G43.5 (Type II)		simultaneous 5-axis
5-Axis Tilted Work Plane	G68.2 (with variations)	CYCLE800	For 3+2 axis positioning
Cancel Work Plane	G69	End of CYCLE800 or TRAFOOF	Cancels tilted plane/transformation

Detailed Comparison by Category

1. Corner Treatment Commands

Rounding

- **Fanuc**: ,R.25 (comma required, optional feature)
  - Example: G01 X50. Y50. ,R.25
  - Works only with G01 (linear) moves
- **Siemens**: RND=25 (standard feature)
  - Example: G1 X50. RND=20
  - Works with linear and circular combinations

- Modal option: RNDM=20 (applies to all subsequent corners)
- Separate feed rate: RND=20 FRC=50

Chamfering

- **Fanuc**: ,C.2 (comma required, optional feature)
  - Example: G01 X4.01 ,C.2
  - Alternative: I, K methods (complicated, 90° only)
- **Siemens**: CHR=2.0 or CHF=2.0
  - CHR = chamfer width in direction of motion
  - CHF = chamfer width perpendicular to motion
  - Example: G1 X... Z... CHR=2.0 FRC=100

2. Unit Selection

Fanuc

- **G20** = Inch programming
- **G21** = Metric programming
- Performs decimal shift, not true conversion
- Should not switch mid-program without changing coordinates

Siemens

- **G70/G71** = Basic unit conversion (does NOT affect feed rate)
    - G70 = Inch dimensions
    - G71 = Metric dimensions
- **G700/G710** = Extended conversion (includes feed rate)
    - G700 = Inch (including feed rate)
    - G710 = Metric (including feed rate)
- Can mix units within a program
- True conversion, not just decimal shift

3. Macro/Parametric Programming

Fanuc - # Variables

```
#1 = 100              ; Assign value
G01 X#1 F300          ; Use variable
#2 = #1 + 50          ; Arithmetic
G00 X[#1+#2]          ; Expression in brackets
```

Variable Ranges:

- #0 = Empty (always null)
- #1-#33 = Local variables (temporary)

- #100-#199 = Common variables (cleared on power off)
- #500-#999 = Permanent common variables
- #1000+ = System variables

Siemens - R Parameters

```
R1 = 100              ; Assign value
G1 X50. F=R1          ; Use R parameter
R1 = R1 - 10          ; Arithmetic
R10 = R5 * 3.1419     ; Complex calculations
```

Variable Range:

- R0-R999 = Available parameters
- Additional system variables with $ prefix ($M, $S, $T, $P, $R)

4. Subprogram Structure

Fanuc

```
O0001 (Main Program)
G00 X0 Y0
M98 P100 L2   ; Call O100 twice
M30
```

```
O0100 (Subprogram)

G01 X10. F100

Y10.

M99         ; Return
```

Siemens

```
; Main Program

G00 X0 Y0

L100        ; Call subprogram L100

; or

SPIGOT_2    ; Call by name

M30


; Subprogram L100

G01 X10. F100

Y10.

M17         ; Return
```

5. Angle Programming

Fanuc (Optional Feature)

```
G1 X0.9 Z0.0

Z-0.1 A135.    ; Angle specified, X calculated
```

Z-1.0

- Requires "Direct Drawing Input" or "Auto Any-Angle" option
- Angle reference varies by control

Siemens (Standard Feature)

G1 X50. Z0.

Z-10. ANG=135.   ; Angle specified, X calculated

Z-100.

- Works in G17, G18, G19 planes
- Counterclockwise positive angles
- Standard feature, no option required

6. Transformation Commands (Milling on turn-mill)

Fanuc

Used to convert X/Y motions into cylindrical interpolation for machining on C-axis spindle:

G12.1 / G112

G01 X... C...

...

G13.1 / G113

Siemens

TRANSMIT switches machine into milling mode on a mill-turn machine:

```
TRANSMIT

G1 X... Z... C...

...

TRAFOOF
```

- Used for milling operations on turned parts

7. 5-Axis Machining Commands

Both Fanuc and Siemens have advanced 5-axis machining capabilities, but they use different commands and approaches for Tool Center Point Control (TCP/RTCP) and tilted work planes.

A. Tool Center Point Control (TCP/RTCP)

Tool Center Point Control maintains the tool tip position relative to the workpiece even when rotary axes are moving, enabling smooth simultaneous 5-axis machining.

Fanuc - G43.4 / G43.5

G43.4 - RTCP Type I (Tool orientation via rotary angles)

```
G43.4 H01 Z1.0 F500

G01 X100. Y50. Z20. B45. C30.
```

- Uses explicit rotary axis angles (A, B, C)
- Controller calculates linear axis movements to maintain tool tip position
- Modal command
- **Optional feature** - may not be available on all machines

G43.5 - RTCP Type II (Tool orientation via vectors)

G43.5 H01 Q1

G01 X118.387 Y-257.363 Z-13.213 I=-0.087 J=-0.995 K=-0.032

- Uses I, J, K directional vectors instead of rotary angles
- More precise with 7 decimal places (requires IS-E increment system)
- Eliminates post-processor calculation errors
- Better for freeform surface machining
- **Optional feature** - requires specific control capability

Key Characteristics:

- Shifts machine kinematics burden from CAM system to CNC control
- Enables programming at tool tip position relative to workpiece

- Works with Head/Head, Table/Table, and Head/Table configurations
- Coordinate system variants: Workpiece coordinates or Table coordinates
- Cancel with G49

Siemens - TRAORI

TRAORI - Dynamic Orientation Transformation

TRAORI(1)

G1 X131.063 Y-257.061 Z-28.170 A3=-.088 B3=-.990 C3=-.106

TRAFOOF

- Uses A3, B3, C3 for tool orientation vectors (directional cosines)
- Similar concept to Fanuc G43.5 but different syntax
- **Standard feature** in SINUMERIK systems
- Can be called within CYCLE832 for high-speed settings
- More integrated with other Siemens transformation features

Alternative: TRAORI_DYN and TRAORI_STAT

- TRAORI_DYN: Dynamic transformation for simultaneous 5-axis

- TRAORI_STAT: Static transformation for 3+2 positioning

Key Characteristics:

- Automatically accounts for tool length in all rotary configurations
- Maintains tool control point relative to programmed workpiece point
- Built-in as standard feature
- Deactivate with TRAFOOF

B. Tilted Work Planes (3+2 Positioning)

Tilted work planes allow programming standard 3-axis operations (with cycles, circular interpolation, cutter comp) in a rotated coordinate system.

Fanuc - G68.2

G68.2 - Tilted Working Plane (Absolute mode)

G68.2 X0 Y0 Z0 I45. J30. K0 Q312

G53.1

(... program 3-axis operations as if in XY plane ...)

G69

Definition Methods (P parameter):

- **P0** (or omitted): Euler Angles - I J K (Alpha, Beta, Gamma)

- **P1**: Roll-Pitch-Yaw angles
- **P2**: 3 points in a plane
- **P3**: 2 vectors defining the plane
- **P4**: Projection vectors

Associated Commands:

- **G53.1**: Tool Axis Direction Control - automatically positions rotary axes perpendicular to tilted plane
- **G68.3**: Define plane from current tool orientation
- **G68.4**: Incremental mode (G91 equivalent)
- **G69**: Cancel tilted work plane

Syntax Components:

- X Y Z: Origin of tilted work plane
- I J K: Rotation angles (method depends on P value)
- Q: Axis rotation order (e.g., Q312 = C-axis first, then A-axis)

Key Characteristics:

- Enables use of canned cycles in rotated planes
- Allows circular interpolation and cutter compensation

- Must call G69 before defining new work plane
- Works with G54-G59 work coordinate systems
- **Optional feature** on many Fanuc controls

Siemens - CYCLE800

CYCLE800 - Swiveling/Coordinate System Rotation

CYCLE800(1,"NAME",0,57,0,0,0,13.064,7.435,0,0,0,0,1,1
)

(... program 3-axis operations ...)

; CYCLE800 ends automatically or deactivate transformation

Parameters (typical usage):

1. Mode (0=off, 1=activate)
2. Frame name
3. Rotation mode
4. Calculation mode (Euler angle order) 5-7. Rotation angles 8-10. Translation coordinates 11-13. Additional parameters 14-15. Options

Key Characteristics:

- More comprehensive than G68.2
- Integrates with Siemens frame system
- Used for both 3+2 positioning and coordinate transformations

- Can be combined with TRAORI for true 5-axis
- **Standard feature** in SINUMERIK systems
- Typically used for positioning rotary axes, not true simultaneous 5-axis

Interaction with TRAORI:

- CYCLE800 positions the work coordinate system
- TRAORI enables simultaneous 5-axis toolpath execution
- Both can be used together for complex operations

Comparison Summary: 5-Axis Commands

Feature	**Fanuc**	**Siemens**
TCP Command	G43.4 (angles) G43.5 (vectors)	TRAORI
Vector Format	I, J, K	A3, B3, C3
Availability	Optional feature	Standard feature
Tilted Plane	G68.2 (+ variations)	CYCLE800
Plane Definition	5 methods via P0-P4	Integrated in CYCLE800

Feature	Fanuc	Siemens
Cancel Plane	G69	End of CYCLE800
Cancel TCP	G49	TRAFOOF
Axis Positioning	G53.1 (with G68.2)	Built into CYCLE800
Use Case - Simultaneous	G43.4/G43.5	TRAORI
Use Case - 3+2	G68.2 + G53.1	CYCLE800

Practical Application Examples

Fanuc 3+2 Operation:

```
O0001

G90 G54

G0 X0 Y0 Z50.

G68.2 X0 Y0 Z0 I45. J0 K0 Q312  ; Define tilted plane

G53.1                   ; Position rotary axes

G0 Z5.                  ; Approach in local Z

G81 R2. Z-10. F100              ; Drill cycle in tilted plane

X10. Y10.

X20. Y20.

G80
```

```
G69                    ; Cancel tilted plane

G53 Z0

M30
```

Fanuc Simultaneous 5-Axis:

```
O0002

G90 G54

G43.5 H01 Q1              ; Activate RTCP Type II

G1 X50. Y50. Z20. I0 J0 K-1 F1000

G1 X60. Y60. Z15. I0.707 J0 K-0.707

G49                    ; Cancel TCP

M30
```

Siemens 3+2 Operation:

```
; Main program

G54

CYCLE800(1,"",0,57,0,0,45,0,0,0,0,0,0,1,1) ; Swivel to 45°

G0 X0 Y0 Z50

CYCLE81(2,0,2,-10,,,100)      ; Drill cycle

X10 Y10

X20 Y20
```

```
CYCLE800()              ; Deactivate

M30
```

Siemens Simultaneous 5-Axis:

```
; Main program

G54

TRAORI(1)               ; Activate RTCP

G1 X50 Y50 Z20 A3=0 B3=0 C3=-1 F1000

G1 X60 Y60 Z15 A3=0.707 B3=0 C3=-0.707

TRAFOOF                 ; Deactivate RTCP

M30
```

Programming Philosophy Differences

Fanuc

- **Approach**: Traditional ISO G-code based
- **Learning Curve**: Simpler for beginners
- **Flexibility**: Less flexible, more rigid structure
- **Features**: Many advanced features are optional add-ons
- **Macro System**: Requires purchasing options for full capability

Siemens

- **Approach**: Structured language with cycles and dialogues
- **Learning Curve**: More complex initially, more powerful long-term
- **Flexibility**: High flexibility, supports mixed programming
- **Features**: Most features built-in as standard
- **Macro System**: Comprehensive macro system included

Key Takeaways

1. **Syntax**: Fanuc tends to use single letters with commas (,C, ,R), while Siemens uses explicit command names (CHR=, RND=)
2. **Features**: Many Fanuc advanced features are optional, while Siemens includes them as standard
3. **Variables**: Fanuc uses # for variables, Siemens uses R
4. **Subprograms**: Fanuc uses M-codes (M98/M99), Siemens uses direct addressing (L)
5. **Units**: Both systems handle unit conversion, but Siemens offers more granular control

6. **Compatibility**: Programs generally need modification when switching between controllers due to these syntax differences

Important Notes

- **ISO 6983 Standard**: Both controllers follow ISO 6983 as a baseline, but each implements proprietary extensions
- **Not Interchangeable**: G-code programs are NOT directly portable between Fanuc and Siemens without modification
- **Control Versions**: Specific features and syntax may vary between different versions of each controller
- **Documentation**: Always consult your specific controller's programming manual for exact syntax and available features

May I ask you a small favor

I would love to know your thoughts on this book. I would be all ears to your feedback because I am eager to find ways to improve. I will listen to negative comments with an open mind and try to fix your concern. Drop me an email at rajeev.cncexpert@gmail.com.

I look forward to hearing from you. Also, if you liked this book, don't forget to leave a review. It will help the world become a more productive place

Disclaimer

The information presented in this book is based on the author's knowledge, research, and industry experience. While every effort has been made to ensure accuracy, the content is provided for educational and general guidance purposes only and is not a substitute for professional engineering or technical advice.

CNC machining involves complex processes, specialized equipment, and potential safety risks. Readers must exercise their own professional judgment, follow machine-specific manuals and OEM guidance, and comply with applicable industrial standards, workplace procedures, and safety regulations.

Cutting parameters, tool selection, and machining strategies vary by machine, material, tooling, part geometry, setup conditions, and control technology. Always test, verify, and validate parameters and techniques in a safe trial environment before applying them in production.

Neither the author nor the publisher assumes any responsibility for errors, omissions, or damages arising from the use of the information contained in this book. The implementation of any techniques, settings, tools, programs, or practices discussed herein is at the sole discretion and risk of the reader.

www.ingramcontent.com/pod-product-compliance
Lightning Source LLC
LaVergne TN
LVHW090257210926
848480LV00001B/46
* 9 7 9 8 2 3 2 0 7 5 9 5 8 *